The economics of
tropical farm management

The economics of
tropical farm management

J. P. MAKEHAM
Department of Agricultural Economics and Business Management
University of New England
Armidale, NSW 2351, Australia

L. R. MALCOLM
School of Agriculture and Forestry
University of Melbourne
Victoria 3052, Australia

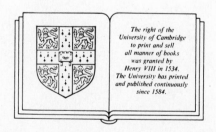

The right of the
University of Cambridge
to print and sell
all manner of books
was granted by
Henry VIII in 1534.
The University has printed
and published continuously
since 1584.

CAMBRIDGE UNIVERSITY PRESS

CAMBRIDGE

LONDON   NEW YORK   NEW ROCHELLE

MELBOURNE   SYDNEY

Published by the Press Syndicate of the University of Cambridge
The Pitt Building, Trumpington Street, Cambridge CB2 1RP
32 East 57th Street, New York, NY 10022, USA
10 Stamford Road, Oakleigh, Melbourne 3166, Australia

First published 1986

Printed in Great Britain at the University Press, Cambridge.

British Library cataloguing in publication data
Makeham, J. P.
The economics of tropical farm management
1. Agriculture – Tropics – Finance   2. Farm
management – Tropics
I. Title   II. Malcolm, L. R.
630′.68′1   S567

Library of Congress cataloguing in publication data
Makeham, J. P.
The economics of tropical farm management
Includes index
1. Farm management – tropics.   I. Malcolm, L. R.
II. Title.
S562.T7M35   1985   630′68′1   85–6720

ISBN 0 521 30894 1 hard covers
ISBN 0 521 31367 8 paperback

SE

Contents

Contents

Contents

Preface

Economic principles, combined with the findings of agricultural technologists, form the basis of the field of study called 'farm management'. From this blend has come tools and techniques for analysing different courses of action and for making decisions. Underlying ways of applying these decision-making aids is the idea of getting a bit more from limited resources.

In this text we have aimed our message at those who directly affect the lives of small farmers in the tropics. This target group includes farm extension and advisory officers, students and teachers of agriculture at colleges and universities, bureaucrats and administrators, employees of the suppliers of credit (and other farm inputs) and managers and staff of parastatals and state farms. Most of these people can use the principles and techniques of farm management economics to help families running either semi-subsistence or commercial farms in the tropics to get more from the limited resources with which they have to work. If the lot of farmers and their families can be improved, it is likely that their society will also benefit.

In Chapter 1 Introduction, we try to show the potential contribution of applied farm management economics to fulfilling some of the more important basic human 'wants', and to highlight the pivotal role, and some of the practical problems, of the extension worker in helping his farmer-clients.

In Chapter 2, Farm management, we have tried to explain our concept of farm management as it applies to both semi-subsistence and semi-commercial farms. Rather than listing a number of general platitudes about 'management', as is common in many management texts, we have been both specific and prescriptive, setting the scene for detailed explanations in later sections.

Chapter 3, Farm analysis and planning, contains a checklist of the main human, physical and financial factors which influence the way in which a farm operates. Each broad category is discussed under three headings: the present situation, potential, and limitations or restraints on achieving the potential.

In Principles of production, Chapter 4, the problem of deciding on the number of weedings a crop should be given to make most profit is used as a simple example to demonstrate some principles which apply to most production situations. This involves explaining some basic economic concepts. These concepts are at the heart of farm management analysis and planning and are used often in the rest of the book. Some readers may wonder what all the fine-sounding terms, techniques and theory in this part have to do with growing a good crop or producing a fat cow. The answer is that, directly, not a lot (of course)! We are not in the business of providing prescriptions on the technical side of crop and animal production. That is the job of the extension service and commercial firms, applied to local situations. Our task is to define the economic concepts and techniques which will help sound farm management decisions to be made, then to show how they can be applied in practice. Detailed applications of the theory are outlined in later chapters.

There are five or six different classes of costs, and several types of returns. It is important that the farm adviser and commercial farm operator should understand the nature of these different classes, as each class has a different effect on farm 'profit'. So, in the fifth chapter, Costs and returns, these are described and explained.

For semi-commercial and larger-scale (private or government) farms it is often necessary to have detailed statements on the financial performance of the operation for each year. In Chapter 6, Farm profits, financial statements and accounts, we show how to prepare such statements. Also, we discuss the nature of 'profit', and show how to keep a few simple records which can be useful.

Chapter 7, Cash flows, contains the main points on how to budget for likely cash receipts and cash payments. There is also a method for working out expected returns and consumption of food on semi-subsistence farms. The method of comparing budgeted with actual results is also explained. Using this control technique lets the farm operator take timely corrective action should the budget not work out as expected.

The techniques for using the concept of the gross margin are introduced in Chapter 8. This is one of the simplest, most commonly used tools for farm financial analysis and planning. Frequent reference is made to it in later sections.

The ninth chapter, Time is money, concerns ways of taking account of the different sums of money which are received and spent at different times, when analysing and planning farm investments.

Next we detail how the techniques and tools of farm management economics are applied to four important areas of farming, viz., how to decide whether it would be worthwhile to make a change in the farming programme, the analysis and planning of farms where annual and perennial crops are grown, economic aspects of animal production and breeding, and choosing the most 'profitable' form of mechanisation, whether it be by tractors, animal power or improved hand tools.

In Chapter 10, Planning changes, there is a detailed explanation of the techniques a farm adviser needs to know so that he can help his farmer-client to decide whether it would be profitable to make a change in his farm programme. We identify two types of change – simple and complex. The technique of partial budgeting is described.

In most tropical countries, cropping is the major farm activity. Chapter 11, Cropping, has details on both the main technical and economic aspects of cropping. Several types of crops are dealt with: short term crops such as vegetables; grain crops which take 4–7 months to grow; those which take 12–15 months (e.g. yams and cassava); and perennial crops (such as orchards, plantations and timber stands). There is also

an explanation of the techniques for analysing and planning farms where a mixture of several crops is grown.

It is usual for semi-subsistence farms in tropical countries to have some animals. In Chapter 12, Animals, we describe the many roles which animals play: producing food and fibre, providing transport and draught, savings and security, tradable assets and (often) social status. The main techniques for the economic analysis and planning of animal activities are described. Special attention is given to the problem of matching animal-feed demands with available feed supply. The biology and economics of animal breeding are also discussed.

In Chapter 13, Mechanisation, the main emphasis is on identifying the costs of acquiring alternative forms of machine service. The three most important forms are ownership, contractors and share farmers. Techniques for deciding which form of service is cheapest are detailed. Questions such as the effect which the annual number of hours used has on costs, when to replace machines, what sized machines to use, and whether to buy new or second-hand, are also covered.

In Chapter 14, Farm development, we show how to plan and appraise a farm development programme – whether it be land development, irrigation, adding a new enterprise such as a poultry unit, or buying of additional land. Great stress is placed on making sure that the physical and technical aspects of the proposed development(s) be fully understood. Then, with an example, we describe the various budgeting and appraisal techniques which need to be used, such as partial budgets, return on extra (marginal) capital, development budgets, discounting and net present values. Most farm development programmes involve both borrowing and risk, and take a period of years before they reach a 'steady state'. It is important that farm advisers and financiers be able to apply the techniques described.

Chapter 15, Farm credit and finance, describes a mixture of theoretical and practical aspects of farm credit. Under 'theory' we consider the financial concepts such as equity, leverage, growth profits, the relation of inflation and interest rates, flat and simple interest, term loans and amortised loans. The theory segment gives a foundation for the practical part of the chapter, which includes a series of checklists (parts of which the financier, adviser and farmer need to consider when discussing a loan).

In the final chapter, Beyond the farm, we raise, but do not attempt to resolve, some important issues which affect farmers but which are decided outside the boundaries of their farms. This chapter contains a miscellany of our thoughts on some broader issues which affect farmers, ranging over topics like the mix of market and planning, marketing boards, credit and mechanisation policies, and training for farm management.

Throughout this text the unit of currency we use is the United States dollar (US$).

Some economic principles and analytical techniques are relevant to small semi-subsistence farmers in the tropics, and can be used to help them to get a bit more from their limited farming resource base.

J. P. MAKEHAM   *February, 1986*
L. R. MALCOLM

Acknowledgements

There are little bits of many people in this book.

We first wish to gratefully acknowledge the help we have received from farmers we have encountered, in several tropical countries, who are trying to do a bit better. Also the experience of many extension workers and applied researchers in these countries was of invaluable help. Thanks, all of you.

In Australia, Alison Affleck and Marion Brown transformed our early drafts into an excellently typed text. Alison produced the final copy – her intelligent, workwomanlike efforts are greatly appreciated. Many thanks, Alison. Marion was the front line worker coping with barely legible messes. A great job, Marion. Andrew Day tackled a similarly daunting task with the diagrams, and skilfully succeeded. Good one, Andrew.

We have been greatly helped by the perceptive insights, as well as the congenial, entertaining company of our professional colleagues at the Universities of New England (NSW) and Melbourne (Vic). In particular, John Dillon, Euan Fleming, Brian Hardaker and Jack Sinden at New England; and Alistair Watson, Neil Sturgess, Alan Lloyd and John Cary at Melbourne. They have done more than their bit, and we owe them more than a beer. Cheers.

As well, Nanette Esparon has been closely involved with all aspects of this project. Thanks, Nanny.

The support given by the staff of the Farm Management Unit of FAO, Rome, especially Neal Carpenter, and also of the Organisation itself, was indispensable. Grazie.

Finally, we would like to acknowledge the invaluable help of the subeditor, Dr Valerie Neal, who prepared the typescript for the printer.

1

Introduction

Farmers have always had to decide what to produce and how to do it. Our aim is to give useful guides to those making decisions relating to farm management, in a form that is easy to understand. We deal with economic principles which underlie farm production and management, and how to apply them. We hope to help the 'doers' and those who help them. Our target audience includes extension workers, teachers and students of agriculture, operators of private farms, managers and planners of state farms, farm development authorities, staff of farm cooperatives, rural bankers and local government planners.

Economic analysis and planning of farm management becomes more relevant as new plant and animal breeds and methods of husbanding them become available, as needs for food and fibre change and grow, and as people's expectations about 'their lot' – of what is and what could be – change (more bread, education, motor cycles, new clothes, and pictures on the wall, too). Farm management economics is simply one way of looking at, interpreting, analysing, thinking and doing something about the situation of farming families and other rural dwellers. There are lots of other frameworks for looking at their world, too. These include politics, religion, history, and anthropology. We will stick mainly to production economics in this volume.

The area of world agricultural production we deal with is 'the tropics'. The tropics lie roughly between the Tropic of Cancer and the Tropic of Capricorn. This area has small seasonal changes in day length, relatively high average temperatures, some areas are wet and hot, and others are dry and hot. Rainfall is plentiful and reliable in some parts, low and uncertain in others. There are 'tropics' in Africa, Latin America, Asia, the Pacific, and parts of Australia. The range of crop products grown includes maize, millet, sorghum, sugar cane, ground nuts, coconuts, beans, rice, bananas, pineapples, coffee, tea, cocoa, vegetables, cassava and yams. Farm animals (though of less importance than crops) include beef cattle, goats, pigs, sheep, oxen, buffaloes, dairy cattle, camels, donkeys, horses, chickens, ducks, elephants, fish – even crocodiles!

Most people in the tropics rely on local agricultural products as their major source of food, fibre and income. In most of these countries, a large proportion of the workforce lives in rural areas, works in agriculture and, relative to the rest of the world, has low levels of income per head, food (and sometimes water) supply, and education. Land-tenure arrangements are complex, sometimes seemingly unjust, and hard to change. Often there is a lack of effective support services to help farm families in such matters as health care, access to credit and to relevant advice on farming methods. Despite some popular beliefs, the tropical resource base usually does not lend itself readily to highly productive crop and animal production. Soils can be poorly structured, low in nutrients and organic matter, and prone to erosion. Attempts to use soils of low fertility successfully can be frustrated by unreliable rainfall, and by pests and diseases. Also, contrary to popular belief, labour can be scarce at critical times.

Furthermore, many parts of the tropics have populations which grow more quickly than the land can feed and service them. The 'creeping deserts' of Africa reflect this problem in its most dramatic form. It also occurs in a less extreme form in parts of Asia and Latin America. Slowing the rate of population growth is vital. We strongly endorse the theme adopted by enlightened population planners who say to their clients: 'Space your children, just like you space plants.

Then you will get better, healthier growth of your seedlings'. Since women do most of the work in growing and husbanding (mothering?) crops in tropical countries, this message is likely to have some impact over the next few decades. There are numerous factors which impinge on the feeding–breeding issue. Making small advances in each part, seems to be the best way of tackling the problem.

Part of the discipline called economics is concerned with looking at how resources are used and what they are used for. One of the reasons for doing so is to see if resources are being used in a good way. One part of using resources in a good way is to use them to meet as many of people's needs for life, and from life – sometimes called people's 'wants'* – as is feasible. Our definition of what the discipline of economics concerns itself with, goes like this:

> Economics deals with a way of thinking about how limited amounts of resources are best used to create the physical commodities and to provide the services which people need. Equally important, it is about the way in which social and economic conditions are created so that people can pursue, with some chance of attaining, their many other life requirements and aspirations.

Two questions worth addressing are: what are people's wants/needs, and of them, to which can the discipline of farm management economics be usefully applied? To us, some of the wants/needs which seem to be part of the human condition are:

1 enough (and sufficient variety of) food and drinkable water;

2 shelter and clothing (these have a range of importance depending on where you live, society's attitudes, your wealth, e.g. fashion clothing);

3 security from violence by others;

4 self esteem;

5 freedom from oppression, freedom of expression;

6 sexual activity;

7 to be physically and mentally active, to be amused, to laugh;

*We recognise that there is a vast literature on the issue of what is a 'want' and what is a 'need'. We have not probed it – it is a study in itself. See, for example, Maslow, A.H., *Motivation and Personality*, Harper & Row (1954).

8 to learn (to become 'better' at something);

9 to have 'more' of things which may make for less misery and more 'happiness' (especially in relation to what some others in the community may have);

10 to do the right thing by god(s) and religious beliefs.

Farm management economics deals quite directly with items 1, 2, 8 and 9, in particular with satisfying more of these wants/needs.

The theory of production economics is often presented in a way that makes it seem unrealistic or of little use in making decisions about farm management. Here we try to provide some usable tools of analysis and planning, based on economic principles, for farming. In the tropics the two main farm situations we will deal with are: (i) where most of the labour, skills and money come from the same household, and most of the production is used at home, with little sold in markets; and (ii) the fully commercial farm, where many inputs are bought and most products are sold. The principles and techniques outlined here apply with different degrees of relevance to each of these types of farming. Most small farmers in developing countries have a semi-subsistence operation – much of the food grown is consumed by the family, but some is sold or exchanged in the market.

Existing farming practices and systems are a result of a mixture of experience, tradition, available resources, the physical environment, levels of technology, and the political, legal, economic and market conditions. There are a thousand good reasons to explain what is presently done, and how and why it is done. Simplistic assumptions about transferring western or eastern bloc economic 'models' to the developing world can be costly to the people and societies which receive such economic advice from evangelical economic 'experts'.

Economists who start talking as soon as they get off the airplane are increasingly being recognised as the burden which they may ultimately pose for the less-developed world. Was it an out-of-touch cynic or a realist who claimed that development aid is a means by which the numerous poor people of the world help the much fewer rich. We have tried to avoid the trap of taking it for granted that 'experts' from countries with highly developed economies have answers to fit the generally unique situations of farm people in developing countries. We have also remembered that trying to make the most money 'profit' (an oft-used yardstick in economics) is rarely the only (and often not the major) objective of farm operators, wherever they may live.

A difficulty when applying economic principles to tropical farming situations is that economics and most other aspects of life of the farm household are all part of the one picture. This is in marked contrast to more developed economies where it is easier to distinguish between production and consumption activities. While in more developed economies, most production and consumption of goods and services pass at some stage through a market, this is often less true of farms in the tropics. Furthermore, production economic theories focus on how the individual, either as a producer or a consumer, behaves. In many tropical situations it is more appropriate to focus on the family group, and sometimes the village. What one villager can do might depend on the wants/needs of all the other people in the village community.

Our focus in this text is on the production, or supply, side of the farming story. This is not to understate the importance of the other half of the story – the distribution and marketing of the production (called the demand side of the economic processes at work). We are not saying, 'well, first grow your product' because production cannot be done properly, or improved, without there also being well developed systems of transport, distribution and marketing. Equally important are the financial institutions and arrangements to ensure that (i) the incentives for production are 'right' and (ii) the facilities necessary for production and marketing of agricultural goods, and for investment in agriculture, are present.

There is some scope for applying some suitably, but sceptically, distilled parts of economic thought, particularly those principles and techniques which are relevant to better using limited resources. Our explanations of the use of farm management techniques are based on the assumption that most farmers are interested in getting a bit more (of something) out of the limited resources they have to work with. When the goal of production is to get more out of available resources, then modern economic principles about making better use of resources are of some relevance.

Resources will be allocated differently, and wealth, power and profits shared in various ways, depending on who owns the means of production; and on different social values and forms of social organisation. However, agriculture is important in the growth of an economy, regardless of which growth theory and ways of applying it you subscribe to. Sound and useful farm production decisions are crucial at the farm level. They may be taken despite, because of, or in complete isolation from, national economic and political considerations and policies. If farmers can get a bit more by using their limited resources and opportunities in a better way, then perhaps others in their society will also benefit.

Target audience

We do not know the details of each local environment in the tropics but we have directed much of our book to local people who know the technical details of particular tropical environments. For example, the extension worker is one target, and the technical side of agricultural production in his local area is just what he knows, and does, best. An extension worker is a person who advises farmers and (often) their wives on farming and family matters. Typically, but certainly not invariably, the person is a male. He is the 'middleman' between rural research groups and farm families. This is his specialisation, his field of expertise, and when we refer to the 'farm adviser', or 'the farmer's adviser', or the 'extension worker', we are saying to him: 'these are some of the important farm management economic aspects of the technical side of your job'.

The extension worker or farm adviser has a pivotal role to play in communicating innovations to farmers, and especially to farm women, who usually do most of the work. The extension worker also acts as a 'troubleshooter' when problems of, say, disease or poor growth of crops, occur. It is likely that he will be numerate and literate, and has had some elementary formal training in general agriculture. It would be silly to expect that most semi-subsistence farmers will be able to work out cash flow budgets, farm financial analyses and loan applications. They will need the help of a trained adviser when dealing with all the problems involving financial analysis, budgeting and planning.

The extension worker has the potential to contribute in a major way to helping farmers in tropical countries to get a bit more from their efforts. Whilst we refer throughout the text to the extension worker helping the farmer, we know that, in most areas, this cannot be done on a one-to-one basis. The extension service is a limited resource, typically spread thinly over a large population of farmers. It is much more likely that the extension officer will be working with groups of farmers, not individuals.

Often extension workers are the 'poor relations' in the bureaucracies administering, and attempting to improve, agricultural development. In the usual situation, the extension worker is relatively underpaid, lacks the elementary facilities for his job (transport such as a bicycle or motorcycle), has to work in isolated areas, often lacks suitable housing, and has poor promotion prospects compared to other public servants and research workers. He also needs to have a special vocation for his job, 'to love his work'. These difficulties are compounded if, as happens frequently in all countries, life in the 'bush' is seen to lack the stimulus and fulfilment of a more urban existence.

So, let us be realistic idealists. We know it will take a long time before policy makers and bureaucrats, with 'empires' to protect, make changes to improve the effectiveness and status of the extension service in many tropical countries. This is partly why students and teachers of agriculture, in both formal and informal learning situations, are important target audiences for our volume. Their potential contribution to their country's agriculture will be realised over the longer term. The other members of our target audience – rural bankers, private farmers, traders and merchants, managers and planners of state farms, executive staff of farm cooperatives, and local and regional government planners – are directly involved and have potential to make immediate and direct impact on the local agriculture. Compared with extension workers, these people's decisions and actions can bring about changes today and tomorrow, rather than next year.

Consequently we hope that bosses in government bureaucracies and private organisations will see merit in sending their staff to relevant short courses based on parts of this volume, to learn about and debate the techniques we have tried to explain.

Tropical farming is complex in terms of biological, social and economic processes. We have tried hard to see the world of the small farmer in the tropics from the viewpoint of small farmers and their advisers. However, since neither of us has ever been small semi-subsistence farmers in the tropics, we acknowledge that we will never be fully plausible. As the saying goes: 'To be a good bullfighter, you must first learn to be a bull'.

Questions

1 Are there any technological developments in crop and animal production in 'the pipeline' which will drastically increase food production in the tropics over the next 25 years? What are they? How will they reach the semi-subsistence farmer?

2 Do programmes aimed at reducing rates of population growth really 'work'? What are the main problems with getting people to accept such programmes? If you can, quote an example of both a successful and an unsuccessful programme.

3 On page 2 we list some human 'wants/needs'. What important ones have we left out?

4 We have placed great emphasis on the potentially key role which competent, 'devoted' extension workers can play in helping farm families in tropical countries to get a 'bit more'. Do you agree with our emphasis? How would you improve the effectiveness of a rural extension service for farmers and their families in the tropics?

2

Farm management

Farm management is about managing farms. Farmers manage farms. Many other people are also interested in how well farms in any country are managed. Governments, extension workers, planners, consumers, bankers, conservationists and politicians are just a few of those who are keenly interested in food and fibre being produced plentifully, efficiently, and consistently.

In most developing countries, farmers get concessions from government in terms of interest below the market rates, subsidised fertiliser, low taxes, and free advice. With existing ratios of prices received and costs paid for farm products and inputs, and rising demand for food due to increasing populations, the combination of (i) concessions and (ii) favourable ratios of prices received to costs paid, means that farming can often be quite a 'profitable' enterprise, using 'profit' in its widest sense (see Chapter 8). (We recognise that governments in many tropical countries keep food prices artificially low; even so, we stand by our claim.)

The size and type of farm may range from a small subsistence plot of less than 1 ha to a state farm comprising all the land of several villages. Farms may be operated by a tenant or an owner, by a manager employed by a cooperative (or state farm), or by an absentee owner. The commonest is the owner-operator, semi-subsistence farm. The same principles of management apply to each type, but of course, with different degrees of emphasis.

We like John L. Dillon's definition of farm management: 'the process by which resources and situations are manipulated by the farm family in trying, with less than full information, to achieve its goals'.

The place of farm management

Two major tasks facing today's farmer in pursuing his and his family's goals are:

(i) how best to incorporate new technology into the farming enterprise
(ii) how to be sufficiently flexible, mentally and financially, to adjust the management of his resources to meet changing costs, prices and varying climatic conditions.

Some schools of farm management thought place a lot of emphasis on record-keeping and accountancy procedures. We do not. We prefer to emphasise the principles of production economics and the technology of farming. In our view, farm management combines the technical and economic aspects of a farm – not forgetting, of course, the human factor (the farm family).

Often, there can be advantages in making use of farm management and economic principles along with new technical methods and capital. There can be a large contrast in profit per hectare between those farms where such ideas are used and those where they are not. The 'management' or 'systems' approach to raising profits has been shown to work in practice in many different farm circumstances.

We recognise that farm management economics is only one of a number of disciplines (e.g. agronomy, animal husbandry, genetics, soil science, engineering, water management) each of which has an important effect on the success of a farm operation. When the reasons for the poor financial performance of a farm are analysed, it is frequently found that:

activities (e.g. crop and animal production) are not being carried out in the best way;

different activities do not go together very well;

more suited or more productive activities are not being tried;

the full potential of the farm resources (human, physical and financial) is not being achieved.

What makes a good farm manager?

We have listed here the knowledge and skills which the operator/manager of a medium-sized, market-oriented farm should have. With modification, these also apply to the semi-subsistence farmer. After all, a farm is a farm, anywhere in the world. So, regardless of farm size, and where they are farming, farmers share many common problems and prospects for increasing 'profits'.

An extension worker needs to be good at communicating with farmers and passing on new knowledge and skills. If his concern is with the 'whole farm' or 'farm management' approach, then he has to have a wide range of knowledge about, and 'feel' for, farming and farmers. The main areas of knowledge which our example farmer should have, or have relatively easy access to, are:

crop production and protection;

animal production;

economic aspects of farm management;

machinery selection and maintenance;

credit and finance;

marketing;

managing labour and communications;

information gathering.

Equally important, he needs the skills to put this knowledge into effective use. To expand on each of the eight 'knowledge' items above, below are listed the skills which farmers should possess with various levels of competence, depending on the relevance of those skills to their situation.

Be able to prepare land; plant, fertilise, weed, water (in some situations) and protect the crop; then harvest, store and market the crop to get the best price, with little waste.

Feed animals properly; prevent disease outbreaks and recognise disease symptoms; achieve high re-production and survival rates; obtain or produce nutritionally correct feed at lowest cost; provide the right housing for effective production, protection, hygiene and harvesting of the animal product.

Use specialist advisers to help analyse the important physical and financial aspects of the farm. Through appropriate records, and other relevant information, be able to work with advisers to produce an annual farm plan, together with budgets, aimed at producing as much food and money as he needs, or as he is able to. Prepare plans of actions in case of abnormal seasons and/or prices. Plan well in advance so that all inputs are available when they are needed, in the correct quantities.

Have harmonious relationships with farm workers by giving them a 'reasonable' amount of responsibility. Be interested in the welfare of people working with him. Know how to establish a clear chain of command so that each person knows to whom he is responsible and so does not have ten bosses telling him what to do. Set up a system of supervision to ensure that the work which is done is of the proper standard. Create a system of communication and involvement, so that all know what progress is being made in achieving the plans and objectives of the farming operation.

Prepare physical and financial reports at regular intervals which are timely, accurate, relevant, brief and clear for the persons who control the farm(s).

Where machinery is involved, be able to choose the most appropriate types for the job. Know the maintenance requirements, and make sure that they are met. Know how to adjust field machines for proper operations and to diagnose and fix *common* faults. Find a good repair centre or mechanic for more complex problems and make arrangements for help in emergencies.

Determine the most favourable forms of credit which can be obtained for different activities. Develop a good, honest working relationship with banker, financier or other credit manager. Be able to prepare a realistic application and finance budget for obtaining credit. Have the ability to know when borrowings are too great to be repaid from the farm income. This involves having, or having access to, budgeting skills.

Assess the different ways of preparing and selling the products. Work out the best way, or combination of ways, of marketing (assembling, preparing, transporting, selling) to give greatest long term benefit.

Be able to obtain relevant information on any problem quickly. Sources of information could be the experience of good farmers, the extension service, private companies, research workers, up-to-date reference books, libraries, friends, college teachers, and agents for machinery, fertiliser and chemical products. An important skill is the ability to assess the value and relevance of bits of information in solving the particular problem in the local situation. Thinking, reasoning skills, combined with common sense and even mini trial-and-error 'experiments' have to be practised and developed.

Most people who are responsible for running farms see farm management as the day-to-day problems of organising and of making decisions about practical and technical matters. Jobs must be done well and on time. It is vital that the mechanical, husbandry and labour operations, which are fundamental to the production of any farm output, be carried out efficiently. A successful farmer should be able to perform the numerous, varied and testing practical tasks, and be able to organise and motivate his labour force. Yet he must be flexible enough to react effectively to changes in expected seasonal conditions, prices and costs.

It is not, however, the purpose of this book to give details on how to grow crops, maintain pastures, manage labour and care for livestock. These can be found in specialised agricultural books, and in leaflets issued by agricultural extension services and farm input suppliers.

Some people regard farming primarily as a way of life, some see it as a way to make a living. Others, and this includes the vast majority of farmers in developing countries, are obliged to consume most of what they grow simply to feed and clothe themselves and their children. Nevertheless, we think that the principles and techniques discussed in this volume can help both those farmers who run a mostly subsistence operation, and those whose farms are mainly commercial, to run them more efficiently, even though the idea of money 'profit' may mean little to some of them.

Resource management

Farmers, in an economic sense, are resource managers who manipulate labour, land, capital and other resources to achieve certain ends. These ends or goals vary with each farmer's responsibilities and sometimes also with his ambitions for himself and his family. Common goals are likely to include: growing enough food of the quality and diversity needed to provide the family with a 'balanced' and interesting diet; security; reducing risk; producing some surplus which can be traded in the market for goods or services which [*or cash crops*] cannot be produced on the farm (e.g. radio, bicycle); avoiding heavy debt; reaching a personal living standard which is on a par with that of peers in the community; acquiring extra land for themselves and for their children; seeing an increase in the value of the assets which they own; reducing drudgery; educating their children; acquiring some savings; improving the appearance of their home and farm; avoiding loss of their farm through (i) bankruptcy or (ii) arbitrary government or private action (e.g. compulsorily acquiring land for, say, a dam and not giving proper compensation).

Generally, farmers are not concerned with pushing production to the extent that they squeeze the last bit of money or kilogram of crop from their land, but most are moderately 'profit' oriented. It is also apparent that most 'traditional' farmers accept relevant new agricultural technology, in whatever form it might take, simply because they become 'better off' somehow, and maybe make more 'profit', by adopting it.

Resources involved in farming

In many tropical countries the land does not belong to an individual farmer but to the tribe, the community, a cooperative or government department or corporation. In these cases, the major resources which a farmer owns are money, his labour, livestock, implements and other farm equipment, and perhaps a house. Where there is a land market, the value of the land owned by the farmer is an additional resource. Often this is a very valuable asset.

The saleable resources which are owned, minus the total of his debts, have a value which is known as the farmer's capital. The farmer who takes a commercial view of the resources he controls will seek to obtain:

a good annual 'profit' from his resources;

increase in the value of his resources, i.e. to make 'capital gains'.

For a subsistence farmer, whose major resource is the family labour force, the first interest is to produce enough food to feed his family with the minimum risk. Second, he would like some cash income. Third, he needs savings which are easily convertible into cash or food in times of adversity. Capital gains are hardly relevant to this situation.

Decision making

Farm management, as a 'formal' discipline, is concerned with helping the farmer to make sound decisions. Decision-making usually involves choosing a course (or courses) of action, from a number of alternatives, which will go towards attaining some of the farmer's goals. The decision-making process has six generally recognised steps:

(i)   having ideas and recognising problems;

(ii) making observations;

(iii) analysing observations and testing alternative solutions to the problem;

(iv) choosing (deciding on) the best course of action;

(v)   acting on the decision;

(vi) taking the responsibility for the decision.

We aim to help the farmer, through his adviser, to make better decisions based on these steps in perceiving problems, collecting information and analysing possible solutions. What should I produce? What method of production should I use? What working capital do I need? Should I borrow? Answering these questions requires an understanding of economic principles, management techniques, finance, and tech-

nology, by the farm adviser. Decisions often have to be made under conditions of change, and uncertainty about the actual outcome (Dillon's 'less than full information').*

Applications to various types of farms

Different types of farmers – the cash-crop farmer, cattle farmer, fruit grower, dairy farmer, or chicken producer – each have their own particular problems. Many farmers combine several different types of production but four basic farm types may be distinguished:

(i)   those where crops are produced by annual cultivation, the produce may be eaten during growth (fodder sorghum), or harvested and sold (sorghum and cotton), or harvested and kept as food (maize) – this includes the shifting cultivator who moves to new land when existing plots lose fertility under continuous cropping;

(ii) those where animals and animal products are produced;

(iii) those concerned with perennial plants (tree fruits such as citrus and bananas, coffee or rubber plantations, perennial seed and hay crops) where there is a flow of income over a long period, and frequently a long time between initial investment and first pay-off, but a need to replant only at relatively long intervals;

(iv) farms with a mixture of several of these activities.

Although the diversity of farm types is wide, the principles and techniques of farm management economics apply, with different degrees of emphasis, to all of them.

Subsistence and semi-subsistence farms

Over large parts of many tropical countries the agricultural activities noted above are carried out on small subsistence and semi-subsistence farms. Many of the principles of management and planning apply, with appropriate modification, to semi-subsistence farms where there is an element of cash income. A few

*Dillon, J.L. (1980). The definition of farm management. *J. Agric. Econ.*, **31**, 257–8.

of these principles apply also to subsistence farms. Below are listed the main features of semi-subsistence farms.

They are generally of small size (area, output).

They do not rely very much on purchased inputs, such as mechanical power or fertilisers.

The farmers need to select crop and animal activities (including hunting and collecting the products of native plants) to supply the basic nutrient needs of the farm household and, where possible, a surplus for barter or sale. Traditional methods of achieving these two ends have been formed by a long process of trial and error and are often quite ingenious. However, increased population pressure has forced many tropical farmers to modify their traditional forms of land use, e.g. by reducing the period of bush fallow or increasing the proportion of starchy crops at the expense of high protein foods.

There is understandable reluctance to try new practices which involve more risk although they may offer prospects of more and/or better food and a higher cash income. For the subsistence farmer, the failure of one of the activities upon which he depends for a significant share of his household's diet can be calamitous. The small- or medium-sized commercial farmer can often afford to take more risks and so be more innovative.

Because traditional activities also carry risks, the semi-subsistence farmer tends not to put his savings into long-term, fixed-capital investments such as buildings and land improvement (even so, he is often prepared to invest in plantation crops which also are long term). He cannot turn many long-term investments into cash in the event of bad seasons and/or prices. He prefers to keep his savings in 'liquid' assets such as cattle, precious metals, rugs and jewellery. These can be converted readily into cash.

There is a relatively low proportion of cash sales to total 'income' of the farm household. The total or gross 'income' has four components: cash sales; cash income from off-farm employment; the market value of farm products consumed by the household; and payment in kind, e.g. food received for services performed. An example of the breakdown of 'income' for a semi-subsistence tropical farm household is as follows:

	Income ($)	
Cash sales	180 ⎫	260
Cash income from off-farm work	80 ⎭	
Value of home-consumed product	410 ⎫	440
Payments in kind	30 ⎭	
Total 'income' for household		700

The main category of direct farm inputs is labour. Cash expenditure on direct farm inputs is small. In addition to payment for inputs such as fertilisers, pesticides, tools, contract services, and interest on loans, cash is also used for overhead and personal expenses. These will include schooling, rent, taxes, and possibly savings and investment. For example, from $260 cash income, $100 may be spent on farm inputs, leaving $160 for meeting overheads, personal expenses, investments and savings.

The other type of expenditure incurred by many such households is 'payment in kind'; where farm products are exchanged for services rendered, or for taxes, or as a tribute to community leaders.

Problems and decisions

Some of the main types of decisions and problems facing semi-subsistence households which can be helped by applying the techniques of farm management economics are how to:

use limited resources such as family labour, outside labour, borrowed money and cash to best effect;

plan and budget expected food supply and cash income for the maximum benefit of the household;

plan in advance alternative courses of action in the event of the budgeted food and cash plan not working out as expected;

select the best combination of activities, both on and off the farm, to produce the food supply and cash income needed to cover essential household needs;

choose between risky alternatives;

plan a crop rotation or intercropping system;

minimise the chance of being seriously harmed by adverse seasons or prices;

estimate available resources to overcome problems of feeding the family an adequate and varied diet; shortages of labour for clearing, sowing, or harvesting; and lack of feed for animals during the dry season of the year.

Human relations in farm management

In the following chapters, considerable emphasis will be given to analysis, budgeting and planning. Even so, we want to stress how important it is that there be effective communication between the farmer, his family and the workforce. Lack of proper understanding can stop the best of plans from being carried out properly, or to the fullest extent.

When a farmer uses labour, either hired or from the family, the quality of field operations can be greatly reduced if the workers are dissatisfied or lack motivation. Similarly, if a farmer tries to introduce new activities or projects into the farm, the expected increase in 'profits' may not occur because the labour force does not understand the reasons for its introduction, or because they feel that the 'boss' is going to get most of the benefits, whilst they gain little. Sometimes the labour force does not have the necessary skills to cope with new programmes. This problem requires training, linked with appropriate cash and non-cash rewards.

One way of establishing good labour relations is to ensure that members of the labour force are given responsibility for a particular segment of the work and for making many of the detailed decisions about how it is to be done.

Management and labour

Good management of labour involves recognising hired and family workers as being people with emotions, needs and goals, not just as standard inputs into a production process. Achieving a contented productive labour force can be traced to the following features of personnel management.

Leadership Involves decision-making ability, sound technical farming knowledge, being well-organised personally and being more an 'encourager' than a 'demander'.

Delegating authority and responsibility This is a key factor in motivating any worker because it develops his/her personal interest. It often creates a sense of achievement and it also allows greater job satisfaction.

Being tolerant but not lax It is important to give credit where it is due and to be diplomatic in correcting mistakes and teaching new skills.

Getting good two-way communications This is based on the ability to see the other's point of view. Good labour and family managers discuss work progress and future programmes with everyone involved and encourage them to contribute their ideas.

Ensuring clear chain of command Workers stress strongly that they must be responsible to one boss only, since it avoids conflicting instructions being given.

Trying to match a job with a worker's interests and skills where possible This can make work more rewarding – perhaps more enjoyable.

While the above features are all important, this list is not exhaustive. An employer who wishes to have both good labour relations and high productivity must supply his workers with good equipment (in proper working order), provide satisfactory working conditions and – above all else – offer incentive through rewards.

Questions

1 Can you improve on John L. Dillon's definition of farm management, viz., 'the process by which resources and situations are *manipulated* by the farm family in trying, *with less than full information*, to achieve their goals'? If so, what is your definition?

2 How many operators/managers of medium-sized commercial farms that you know have the knowledge and skills listed on pp. 6–7? From where can they be obtained if needed?

3 Does Dillon's definition of farm management have any relevance for advisers of semi-subsistence and subsistence farmers, whose features are described on pp. 8–9?

4 In what situations is having good human relations in farm management important in tropical agriculture?

5 With which of the points we have made in this chapter do you disagree? What would you have said?

6 What factors about farm management have been overlooked?

3

Farm analysis and planning

This volume is about using some of the useful techniques which apply 'economic' ways of thinking to farm management. Farming is mostly a physical, biological activity – but it does have its human, as well as its economic aspects, which cannot be ignored. We will ask you to pretend that you are to go out to a farm that you have never seen before, and to bring back a report on what the farm is like. Some of the questions you will have to answer are, what:

are its strengths and weaknesses?

makes the farmer and the farmer's family 'tick'?

does the farmer and his family need from the farm?

does he (they) want to do with, or to have, or to get from, his (their) life?

are the main problems of the farm?

is the financial picture, the limitations, and the potential?

We have prepared a list of the questions that you would need to find answers to, if you were going to report back on the farmer and the farm. Remember, we are approaching this subject from an 'economic' point of view, so you will need to tell us how you are going to advise the farmer how to make more 'profit'. We will define 'profit' and 'economic point of view' later.

We have tried to cover most aspects of farm analysis and planning. Not all of the topics covered in Table 3.1 will apply to your situation, and there will be some that do apply which we have not mentioned. After studying Table 3.1 you should use it as a guide to compile your own list for your situation. Table 3.1 covers human, physical and financial aspects. We expand on these three sections in this chapter but it is important to note that for a full explanation, you will need to read the relevant chapter. In economics, unlike much of science, many of the terms used can mean different things to different people. For this reason, in later chapters we discuss the ranges of meanings and interpretations commonly given to key terms and ideas used in economic analysis.

From this part of the chapter on, we expand on the topics mentioned in point form in Table 3.1. Each part of this table has three components:

factors to be considered when analysing (or knowing) the present situation on the farm;

the potential for improving the existing situation through better technology and management and exploiting some possible options

things (or factors) which could restrain (prevent) the potential from being fully achieved.

This three-part breakdown of each segment is repeated until the end of the chapter.

Human

Farmers' goals

Knowing the farmer's goals for himself and his family will help to explain why the farm is being managed in the way it is, and how (if at all) the farmer might be able to exploit the potential of the farm. Farmers' goals vary widely (see p. 00). When the farmer's goals are being considered in the light of the potential of the farm, it is essential to determine whether the resources of land, credit and skills are compatible with these goals. If they are not, then the goals may have to be modified or redefined.

Sometimes, the farmer may not be fully aware of the possibilities for exploiting the full potential of the

Table 3.1. *Key factors in farm analysis and planning*

Part (A) – human

Present	Potential	Some restraints on achieving potential
Goals of farm family	Scope for change of goals	Lack of knowledge of opportunities Resource base limited Content with present situation
Skills of farm family	Prospect of improving skills	No informal or formal ways of acquiring more skills

Farm family labour supply
    Numbers
    Age
    Sex
	Better use of family labour through diversification into both rural and non-rural activities, or through some changes in who does what, when, how	Small scope for extra enterprise diversification
		Knowledge and skills for non-rural activities difficult to acquire
		Traditional labour roles for different members of the family

Part (B) – physical

Present	Potential	Some restraints on achieving potential
Cropping		
Type of crop Short term (3–4 months) Medium term (4–7 months) Year long (12–14 months) Perennial (2–50 years) Cultivation practices for each type of crop	New crops Change in mixture of crops More land clearing Improved cultivation practices	Uncertainty of performance of new crops New mixture of crops may not satisfy needs for diversified diet Availability of land for clearing Cost and difficulty of clearing New crops may not fit in with the essential crops which have to be grown No suitable, improved cultivation
Cropping system Cropping intensity ('R' Value) Shifting or 'fallow' Sole crops Double cropping Rotations Intercropping Interplanting Mixed cropping Relay cropping	Increased cropping intensity More intercropping Double cropping Interplanting	Rainfall inadequate Labour needs Relevant technology lacking Risk from increasing cropping intensity
Methods of maintaining soil fertility Fallow Legume-based rotations Non-legume rotations Chemical and mechanical means Mixture of the above methods	Increasing legume crop percentage in the rotation Using longer or better fallows More effective use of crop residues and stubbles to maintain structure More 'chemicals', draught animals or 'machines', e.g. sprayers	Vigorous growing legume species are not available and/or successful or useful Pressure of population growth on limited supply of land Stubbles sold as straw and not returned to soil or fed to grazing animals

		Limited supply of organic matter to maintain fertility over the whole farm Supply, cost and lack of knowledge of fertiliser and sprays

Inputs used

Planting materials, type and amount Availability and cost	Improved or cheaper types available	No suitable alternative plant, materials or types

Cropping labour needs

Main operations Amounts Skilled or unskilled How seasonal peak demands are met (share, hire, exchange)	Developing cropping systems which reduce peak labour demand Improving arrangements for obtaining labour at peak times; labour saving, e.g. chemicals, machines	No feasible alternative cropping system available High wage rates for casual labour at peak times; no exchange labour available

Crop product utilisation and marketing

Marketing channels Sales/domestic use On-farm processing Storage/storage loss	Improving presentation or marketing arrangements for product Reducing storage losses	No facilities available to improve presentation or marketing channels Cost of improved storage too great; too many parasites; no suitable 'chemicals' or structures readily available to reduce losses

Irrigation

System of irrigation How limited water is used among competing activities	Irrigation development Better use of water Improved drainage and recycling of water	Lack of capital Lack of water Poor water distribution systems Inadequate knowledge

Feed supply for animals

From pasture and crops

Proportion of non-edible vegetation Pasture species and crop residues Fodder crops for animals Expected seasonal (wet and dry) feed production from pasture and crop Variability in seasonal pasture production Availability and use of feed from agistment, leasing and migration	Increasing feed supply by better pasture, fertilisers and fodder crops Growing more trees and shrubs which are edible Improving strategies to reduce losses in bad seasons and increase profit in good seasons	Cost of fertiliser, pasture seed and growing fodder crops, relative to value of production which would result from animals which consumed the extra feed Lack of information on livestock and feed prices Lack of capital to finance fodder crop/pasture improvement, or to buy the stock which are needed (even though, if it could be done, it would be profitable)

From grain and purchased feed from factories

If feed mixed on the farm, what are sources of the grains etc. used How is the best ration decided How is the feed mixed, stored and fed to animals If feed bought already mixed from the	Improving quality of farm mixed rations by getting better advice on nutrition value of feeds and needs of livestock Reducing wastage in storage and feeding	Lack of relevant information from local extension workers Lack of suitable equipment for storage and feeding, or lack of funds for equipment

factory, is *supply* and *quality* reliable
Performance of animals on factory
feed versus farm mixed feed

Independent testing by extension
service of feeds supplied by factory

Animal activities (converting feed)

Type of animals

Cattle, goats, sheep, camels, chickens, ducks, pigs, horses, donkeys, rabbits, fish, elephants, other	Introducing other species	Climate and environment Disease risk Cultural and religious factors Costs and risks of change

Structure of herd, flock or group

Numbers Proportion of females to males Age of various classes of stock Source of semen (natural or artificial insemination)	Increasing numbers and proportion of females Reducing average age of group Specialising in a certain type or class of animal	Low production rates Poor nutrition No artificial insemination readily available

How system is replaced

Self-replacing Bought-in replacements	Increasing reproduction rates Acquiring better quality bought-in replacement stock	Lack of knowledge on methods of increasing reproduction rates Unavailability of suitable bought-in replacement stock Preference for own bred stock

System of feeding

Grazing, crop and crop residues, hay, agistment, home-mixed or purchased concentrate feed, feed supplements in dry season	Improving utilisation of feed supplied Increased use of feed supplements in dry season	Lack of knowledge Inadequate facilities Price and/or availability of supplements

Husbandry

Labour for care of stock Housing, yards, veterinary chemicals and medicines Season of birth of young	Reducing disease and parasites Improving housing and yards to improve efficiency of disease control and handling Change time of birth of young	Lack of competent and trustworthy people to care for animals Cost and/or unavailability of suitable veterinary services and medicines Absence of appropriate designs and/or materials for housing, yards, and animal-handling facilities

Genetic quality

Methods used to maintain or improve quality	Introduce improved strains Use of artificial insemination Cross-breeding for hybrid vigour	Outdated ideas on selection methods Lack of suitable new strains adapted to environment Prejudice against cross-bred animals Customs opposed to artificial insemination No artificial insemination services available

Machine and animal power

How machinery[a] services acquired

Owned Contract Share farm Exchange Leased	Acquiring machinery services which do a more *timely*, better and/or *cheaper* job Improving existing *simple* hand tools and equipment	Capital costs of investing in own machinery Availability of *reliable* contract services (either from private, co-operative or local government sources) No local research by agricultural engineers on improving simple tools and equipment at present in use

Age and value

Number of years when it has to be replaced Expected cost of replacement Adequacy for the job Adequacy of maintenance	Increasing life of machine through better maintenance Using better designed machines	Lack of semi-skilled labour or lack of spare parts to maintain machines

Animal power

Type

Ox, buffalo, donkey, horse, elephant, other	Role of different species Use of improved designs of animal powered equipment Better husbandry and nutrition to improve work output and life of animals	No suitable, adapted, alternative species Skills, knowledge, nutrients and medicine needed to improve husbandry and feed lacking

How acquired

Own Hire Contract Share Exchange	Improving timeliness and quality of job done by various sources of animal power services	No alternatives to ownership

How maintained

System of replacement Method of feeding and housing	Using improved husbandry, quality of stock and training to get better work and sale value	Lack of good training skills and facilities, and quality replacement stock

[a] This includes hand tools.

Part (C) – financial

Present	Potential	Some restraints on achieving potential
Total capital value of farm, stock and plant	Expected future returns to farm resources	Poor management No relevant new technologies to

Extent to which inputs are purchased and output is sold, in markets or through cooperatives
Household consumption of saleable products
Activity gross margins
Reasons for level of gross margins
Levels of overheads, which includes essential household payments
Total gross margins minus overheads
Debt levels and terms of loans
Return on resources
Structure of 'business' e.g. family or group holding, partnership, small company
Debt compared to equity
Farm cash surplus
Farm 'profit'
Liquidity (i.e. whether some resources are readily saleable e.g. cattle, jewellery, carpets)

Increasing gross margins of activities
Best mix of activities
Introduce new activity(ies)
Expected rate of return on extra capital investment in new or expanded activities
Avenues for extending terms of loans, getting new loans
Possibility for changing structure
Prospects for increase in land value (if a market for land exists)

increase gross margins
Credit very limited and not geared to realities of farm production or development cycle
Tribal or family customs prevent change in structure of ownership and/or management
No advice available to help work out investment alternatives

---

resources which he and his family operates. Other farmers might be content with their lot, and do not have high hopes and wants (expectations) which cannot be achieved. 'Too high' expectations can make you angry or disappointed when they are not fulfilled.

Skills of farm family

Identifying where the farmer and his family's strengths, weaknesses and preferences lie in doing the various physical and mental tasks, helps to explain why the farm runs as it does. Importantly, if a change is being thought about, it is vital that the farmer should have the necessary skills, knowledge and personal make-up to handle the new, maybe more intensive and technically complex, situation.

Farm family labour supply

The numbers, age and skills of the farm family have to be taken into account in any analysis of the farm. Of particular interest are the methods used to meet peak work loads and the skills available to do such specialised tasks as basic maintenance and repairs to machines, pumps and generators, or the care of sick or uneconomic animals.

Possibly, the numbers and skills of the farm family could be used more productively if the farm were more diversified, and if sideline activities were introduced. Perhaps, also, there is potential for engaging in some non-rural activities such as small trading, handicrafts, sewing clothes, brick-making and transport. It is not always possible to diversify very much, but small non-farm activities are often feasible. Some off-farm activities require a moderate amount of skill, and possibly working capital to buy tools and equipment (e.g. for repairing motorcycles or making clothes). These skills and funds are usually hard to acquire (but it is not always impossible) for the young, motivated members of a farming family.

Other, simpler non-farm activities do not need a great deal of knowledge and skills. In these situations, it is important to assess whether there is a worthwhile market for the goods or services offered; there is unlikely to be a problem in terms of a shortage of labour or skills.

Physical

Cropping

Types of crops

The main points which need to be examined when analysing the cropping enterprise on a farm are the types of crops grown, whether they are annual crops such as maize, or tree crops which grow for many years (e.g. oil palms). Within this range, there are short-term (3–4 month) crops such as tomatoes, year-long crops like cassava and yams, and crops which often take 3–4 years to reach full maturity. It is also necessary to know what cropping system is being followed (sole, double, multiple, mixed, relay, strip and intercropping; these terms are defined on p. 96. It is also vital to find out the methods of preparation, planting, husbandry and marketing used for each crop.

When evaluating the potential for the crop enterprise, a number of things need to be looked at. These include: the scope for new crops; changing the mixture of crops; having better rotations and intensifying cultivation (say through more double-cropping and more land clearing).

Factors which could restrain (or prevent) the farm achieving its full potential could be uncertainty about how well a new crop grows, or tastes; the crop's market; the labour requirements; the new mixture of crops not satisfying the full needs of the family for a balanced, diversified diet; no suitable land for clearing, or the cost and/or difficulty of clearing bush being so great that it would not be worthwhile to plant the new crop. The latter situation is not very common today in areas where there is great population pressure on any land which is suitable for cropping.

The main limitations on more intensive land-use through mixed, double and intercropping are soil structure and fertility, shortage of rainfall at critical times, labour, and lack of relevant information.

Methods of maintaining soil fertility

This is a vital part of any analysis of a cropping farm. Indeed, this issue is at the core of sound land-use policies and programmes. It is of particular urgency and importance in developing countries because of the high and growing pressure of population on suitable cropping land. Most farmers are finding it necessary to increase the level of cropping intensity. This brings a big problem – declining soil fertility. For a fuller discussion on soil fertility, rotations and intercropping, see Chapter 11.

A 'fertile' soil is one which has adequate plant nutrients, has good structure, is relatively free from weeds and is not eroded. It should also have a high population of 'useful' microorganisms such as earthworms and nitrogen fixing bacteria, and a low population of harmful ones e.g. certain fungi, nematodes and viruses. The main ways of maintaining and/or improving soil fertility are: bush fallow; conventional fallow; having a rotation which includes legumes (e.g. groundnuts, beans, peas, clover and alfalfa); rotating different crops so that diseases or pests specific to one crop do not build up; adding nutrients through fertilisers; killing weeds by hand or with sprays; and maintaining structure by restoring 'humus' to the soil. The latter usually involves some system (manual or mechanical) of putting crop stubbles back into the land, or using animals to eat the stubble and then partly return it through their dung and urine.

The other way of maintaining soil fertility is to spread on the soil organic matter such as household wastes and animal manure collected close to the house. Generally, there is not enough of this organic matter to replace *all* of the nutrients which have been taken from the fields by crops. That is why 'artificial' or 'mineral' fertilisers such as superphosphate, urea, ammonium nitrate and more complete 'mixed' or 'compound' fertilisers are widely used throughout the world.

Having looked at what is currently being done on the farm about soil fertility – whether it is being 'plundered', maintained or improved – it is then necessary to see what potential there is for improving the situation. Some questions to be answered are:

Can more bush or non-bush fallow be introduced into the cropping system?

Can more, or better, legume crops be grown?

Can more crop residues and stubbles be returned to the soil either directly or through grazing animals?

What would happen if more cash crops and fewer legume crops were put into the rotation?

How can 'fertility' be improved by using more chemical fertilisers and sprays in a better way?

Rather than concentrating all the organic matter from household waste and animal manure (chicken, goat, pig, sheep, etc.) on small areas close to the house or compound, could some of it be spread on the more distant cropping fields?

Is best use being made of the organic manure which is spread on gardens close to the house?

Could more grazing animals be introduced into the land-use system?

We will now discuss some of the things which might *act against* (restrain) the achievement of farm potential. A major problem is that of too many people competing for too little land, as there is a limit on suitable land supply, and a limit to how much better technology and management can increase yields. Thus, the rate of increase in the number of people needing to use some of the 'fruits' of the land must be reduced if the useful available land is not to be 'plundered'. The psychological and physiological aspects of family planning are relevant here.

These problems are all linked. The increasing market value of straw for building, fencing and fuel, makes it less attractive for farmers to return it to the soil or let animals graze it in the field. The need for firewood leads to the removal of trees, denuding the land and so making it prone to erosion. Of course, straw is used to make 'hay', brought home from the field and fed to animals in small barns, shelters and sheds. Too often, however, the resultant manure is wasted and is stench-making rather than soil-making (i.e. soil-building).

Most farmers now recognise that fertilisers 'work'. Restraints on their use are usually lack of supply and sometimes (but rarely) cost, where it is known that they 'work'. With sprays, especially for weeds, there are the problems of lack of knowledge of the finer, (but extremely *vital*), points on how much to use, how to apply them for best results, and how to avoid killing half the crop with them. The long-term effects of *some* chemicals are also unknown. Some of these problems could be overcome without too much effort on the part of the extension service and chemical companies, over a 4–5 year period. The main restraints on using organic fertilisers to improve soil fertility on the whole farm is the small supply compared to the large demand caused by 'fertility-draining' cropping systems.

Labour in cropping
Sometimes, in cropping, there are peaks of demand for workers but not enough available and/or willing hands to do the job on time. There are many possible arrangements for getting the job done. For example, by farmers agreeing to share the labour cost among themselves, i.e. exchange. They can also hire workers at the going rate for cash wages. There are many arrangements between these extremes, e.g. share farming, renting and bonus deals which rely on crop yield and/or price. The potential for improving either the timing of the job or the standard (quality) of operation is worth looking at.

Restrictions on improvements to the complex 'owner–worker' relationship are severe, and difficult to overcome, given the history and structure of arrangements for getting labour, in each local situation.

Irrigation
The system of obtaining and delivering water has to be investigated. Of particular interest is the reliability of the water supply. Water quality (especially salinity and some idea of the content of toxic materials) also needs to be determined.

As water is usually a scarce resource, the way in which it is distributed between the various alternative uses for it – annual crops, tree crops, animals, household – has to be analysed with a view to finding the most effective use for it. Some aspects of exploiting the potential of the limited water supply include reducing wastage, better distribution, giving highest priority to activities which show greatest response to irrigation at times of annual rainfall shortage, improved drainage so that no salting and waterlogging occurs, and the scope for recycling some of the drainage water. Better pumps, piping, spray drips, and better land levelling and contouring, could all justify their cost.

Some restraints on obtaining the full potential are lack of capital either to develop more irrigation land or to buy the equipment to distribute the water properly, unreliable distribution systems from the main source of supply, and lack of knowledge about new developments in irrigation technology.

Feed supply for animals

From pasture and crops

When considering pasture and crops a farm analyst first needs to take account of the topography (whether steep, undulating or flat), the soils, presence of rocks and non-edible vegetation such as trees, thorny shrubs and unpalatable bushes. Next to be assessed are the pasture species present, their palatability and digestibility, what crop residues are available, and when; also, whether special crops are grown just for animal fodder.

Each climatic zone normally has at least one wet season and one dry season. This implies seasonal pasture and crop growth. It is necessary to define what the 'average expected' feed supply will be in each season. In most cases, each year is different, and the actual seasonal feed production might vary from very good to poor. It is important to find out how much variability exists around the 'average' production. Equally important are the strategies which the farmer uses in the event that the seasonal feed production is worse or better than the 'average'. How does he keep his losses down in a bad year, and how does he make the best use of the opportunities offered if a very good season occurs?

Another aspect of feed supply that is often important is whether some of the feed comes from grazing the animals on other peoples' land, for which a fee is paid; the reliability of this source of feed needs to be determined.

Feed for animals can also be obtained by moving the animals hundreds of kilometres each year across the country (or countries), following a good feed supply which occurs at different times of the year in each locality. The owner of the livestock does not necessarily take part in this migration – he pays a fee to skilled herders to care for the stock. This can often be a relatively inexpensive means of obtaining feed, but is becoming less so as more grazing land is used for cropping.

When considering the potential for increasing feed supply from pastures and crops, it is worth looking at whether it is feasible to sow improved pasture species on the land used for grazing, if there is a role for putting fertilizers on the pastures and if it is possible to increase fodder crop production by using improved varieties and some fertilizers.

The main restraints on getting the full potential feed supply are the costs of seed, fertiliser and growing fodder crops, relative to the amount of extra feed that grows as a result of using such techniques, and the use to which the extra feed is put. If lots of extra feed grows, and it is used efficiently by animals whose product (e.g. milk, carcase) bring a high price in the market, then it could be worth doing. If not, it could be a waste of money.

Another restraint can be lack of knowledge about livestock and feed prices in situations where the farmer wishes to exploit the good years and cut his losses if the season is bad. This is not always a serious restraint, as the informal information flow on such topics, even to people who do not read and might not even have radios, is often rapid and accurate.

There are two ways of supplying grain-based feed to animals: mix the grains on the farm, or buy the feed from a processor who has already mixed it. The person studying the existing situation on the farm needs to find out whether the grain used is home-grown or bought, how the best mixture of grains and additives is decided on, and the methods of mixing, storing and feeding. Where feed is bought from a company which mixes feed, the reliability of the feed supply, the quality and the consistency of quality, are important. The relative performance of animals on home-mixed feed and factory-supplied feed, and the relative costs, need to be ascertained.

There may be potential for improving the quality of farm-mixed rations (and hence animal performance) by getting advice from the extension service on nutritional values of various grains and on the specific feed needs of the different classes of animals being fed. In some situations, a feed-analysis service provided by the extension service can help ensure that the right blend of grain and additives is being used. Wastage during storage (through, e.g. insects, rats, or weather damage) and when feeding out, may be important causes of loss which could be prevented or reduced by better structures and simple equipment.

The main barriers to achieving the potential for improvement in grain-based feeding systems are lack of specific knowledge by extension workers about particular animal-production methods and requirements, no readily available feed-testing services, and little suitable storage and feeding equipment. The last is not usually a serious restraint, as ingenious solutions using local materials are frequently found.

Animal activities

When analysing animal activities on the farm, first find out the types of animals being run. Then an inventory of the numbers in each herd, flock or a group has to be made, containing details of the proportion of females to males, age of animals in each class, how many animals are kept for replacements each year, whether the semen supplied during mating comes from natural sources (e.g. bulls) or from an artificial insemination (AI) service, provided either by government or private services.

It is vital to find out how each livestock group replaces itself. There are two ways that replacement occurs: (i) by breeding one's own replacement stock and keeping enough to replace the older animals when they are sold, and (ii) by buying in young replacements, e.g. in broiler chicken activities.

The next step is to examine the systems of feeding – whether it be from grazing pasture, crop residues and crops, or from hay, agistment, home-mixed or purchased concentrate feed, or perhaps using a combination of some or all of the above.

The methods used to maintain or improve the genetic merit of the herd or flock needs to be looked at closely. Identifying and breeding from superior animals can often provide extra income at relatively small cost.

In assessing the potential for improving the animal enterprise as a means of converting stock feed into money, a number of aspects are worth examining. It may be possible to introduce different species or other classes of animal. It is also worth investigating whether it is plausible to increase both overall animal numbers, and the proportion of females to males (usually, females are more 'profitable' than males). Very old animals are usually less efficient than mature but younger ones; if the average age of the group can be reduced there should be a rise in efficiency and profit.

When it comes to exploiting modern genetics to help achieve the potential which may exist, the following options should be considered:

the introduction of improved strains of animals;

the use of artificial insemination;

the use of more crossbred animals to obtain increased hybrid vigour, and disease and pest resistance.

Using limited feed supplies efficiently is a key factor which affects the profitability of animal enterprises. This is the case in both grazing and 'factory farm' production. There are numerous avenues to be explored ensuring efficient use of feed, e.g. more controlled grazing, segregation of age groups, using feeding facilities which reduce wastage and mixing different species of grazing animals (cows, goats, sheep).

Measures taken for disease and parasite control are often not very costly relative to the improvement in animal performance (and profits) which result. The first need is for animals to be properly fed, as poor nutrition makes them much more prone to disease and parasites. Medicines, chemical and veterinary advice are also integral to insect and parasite control.

Some important restraints on the exploitation of the potential for improving animal enterprises include climate, environment, disease risk, culture and religion. For example: sheep do not normally thrive in the wet tropics; cattle are prone to attacks of tse-tse fly in the wetter parts of Africa; Hindus do not eat cattle, nor do Muslims eat pigs.

Though replacement through bought-in replacement stock rather than through a self-replacing system might be better, it may not always be possible to acquire 'superior' quality replacement stock of the type that the farmer wants.

Barriers to using feed more efficiently include not knowing the best way of doing it, and a shortage of proper facilities (sheds, yards, equipment). Often better husbandry practices are not possible because there is a lack of competent people to care for the animals. In the case of grazing animals which are often taken long distances away from home, there is sometimes a problem finding trustworthy people to care for the animals (these situations are the exception rather than the rule).

Sometimes there are no suitable medicines, vaccines or chemicals close at hand. A more limiting restraint is lack of veterinary and animal-husbandry advisory services to help the farmer to choose which modern drugs to use and how best to get and use them. Lack of appropriate designs for housing, yards and handling facilities which the farmer and his adviser can adapt to an individual situation, using local materials, can often be a restraint on achieving better results. 'Appropriate' design is an area which could produce high returns for research investment. The research would start with what the farmers know and do now. It would identify the worst and best features of their present set-up, then consider the principles that are most important, and the methods that have been developed, from the experience of 'developed' countries. Its end result would be to produce designs which embodied the best elements of the experience of both groups.

The main restraints on exploiting genetics to the full are ignorance of the practical implications of genetic research, prejudice against cross-bred animals, religious customs opposed to artificial manipulation of conception, and the absence of AI skills and facilities within easy reach. There is also the problem, where it is desired to introduce new strains (*not* species), of finding those strains which are as well adapted to the local environment as the strains that have been used there for centuries. Too many disasters have occurred when high-yielding European dairy cattle (*Bos taurus*) were introduced to areas where the locally adapted milking strains were thought to be unsatisfactory. Often, most of the 'pampered' European cattle died of disease in a short time, because they were not adapted to the local environment.

Machinery and animal power

Machinery services

When analysing how the farmer uses machinery, the question is not 'what machinery, equipment and tools do you own', rather it is 'from what sources do you get your machinery services'. A 'machine' such as a plough is something used to provide cultivation *services*. These services can come from the farmer's own plough, by borrowing it from a neighbour, by hiring one, or by using contractors to supply a complete 'package' of plough, labour and power services.

Having found out what services are provided by machine, the manner in which these services are provided has to be defined. These services can come from equipment which the farmer owns (by himself, or jointly), from contractors (government or private), from share farmers, through one of the many forms of exchange arrangement with other farmers, in some cases through hiring or leasing, and perhaps from other sources.

For machinery which is owned, singly or jointly, the farm analyst needs to find out how many years it will be before the machine has to be replaced (this is extremely difficult to do with much precision because when a machine is replaced depends upon many factors), what the expected cost of the replacement machine will be (less any trade-in value on the old machine) in the year of replacement, how well it is being maintained, what tools and facilities for maintaining it are needed (and available), and whether it is the right size and design of machine for the range and size of the jobs it has to do. Availability of spare parts and repair services need to be defined.

When appraising the potential for using machinery services, it is necessary to know if such services could be obtained which would do the job better, which would allow operations such as sowing, weeding and harvesting to be performed on time, and be cheaper (e.g. using more contractors or share farmers, reduced ownership of obsolete machinery, the availability of good second hand machinery).

It is often well worth considering the potential for minor modifications (which may bring a major improvement) to the simple hand tools and equipment (hoes, sickles, threshers, sprayers) already in use. There is scope for use of new types of materials to improve the quality of the tool's operation and/or the mobility of the equipment plus the ease of the task. The areas which offer greatest potential for improving 'owned' machinery services are better maintenance to prolong the life of machines; and more correct adjustment of equipment to enable the machine to do its job properly.

The ingenuity of some agricultural engineers and

farmer-inventors results in a continuing flow of simple modifications to existing machines, and better designed machines, onto the market. The progressive extension worker is constantly on the lookout for relevant developments which could help his farmer-clients.

Animal power

Many of the services provided by machines can also be supplied by animals, e.g. ploughing, threshing, pumping water and transport. It is necessary in any farm analysis to note what sources of animal power are being used and how their services are obtained – ownership, sharing, contract, hiring or exchange. Other aspects to be investigated are how the draught animals are replaced – breeding or purchase – and how they are housed and fed in order to have them working efficiently. When estimating the potential for improvement, it is necessary to pose the following queries:

Would a different source of obtaining the services of animal power lead to a better or more timely job (e.g. ownership rather than share, less use of contractors, more hiring from neighbours, or vice versa)?

If the animals were better fed and cared for, would this result in better crops, less human effort and a higher sale value for the animals when they have finished their working lives?

Are better quality draught animals available?

Would more thorough training make the animals easier to work later in the field?

The chief barriers to fulfilling the potential of animal power are that no suitable, adapted alternative species are available; the skills and knowledge to feed and husband the animals may be lacking; proper medicines and top quality feeds often cannot easily be had; possibly there may be no contract or hire services and the only alternative is to own the animals either outright or jointly. There may also be a shortage of traditional animal-training skills as younger people take more interest in machines such as bicycles and motor cycles. They find that the time, effort and patience needed to train work animals makes it an unattractive pastime.

Financial

The first thing to consider in a financial analysis of the present situation of the farm and the farm family is the market value of the resources the farmer uses and/or controls. Resources are usually classified as: livestock; equipment, tools and machinery; cash savings; valuables such as carpets and precious metals; and (where there is a market for land) the value of the land and buildings. The inputs which are bought for use on the farm, the output which is needed for consumption on the farm, and the amount which is usually available to be sold, all need to be determined.

Next, estimate the average gross margin (gross income less variable costs) per unit of land, or labour, or head of livestock, for each activity, e.g. beans, yams, maize. The unit of measurement chosen should be that factor which is most limiting in the farming system being followed – most often it is either per unit of land, or per unit of labour at a limiting time. (Remember: gross income includes saleable food consumed by the farm family.)

The reasons why the gross margins of each activity are at the level which they are at present, should be investigated. Maybe, relative to the average gross margins being obtained by other farmers with similar farms, the figures on this farm are low. This could be due, among other things, to poor management, out-of-date methods or shortages of labour at critical times.

The total overhead costs have a big influence on profitability; finding out what they are is a 'must' in any farm financial analysis. Overheads are costs which must be met each year, whatever activities occur. Overheads consist of such things as:

minimum cash living, schooling and clothing costs for the farm family;

finance costs such as interest and fixed-loan repayments;

taxes and government levies;

social and religious obligations;

costs of running, registering, repairing farm motor vehicles such as cycle or pick-up truck;

annual repairs to buildings, structures (such as fences, dams and pumps);

net costs of replacing tools, machinery and vehicles; wages of permanent (non-family) workers (if any).

There will also be other overhead (fixed) cost items, which are specific to certain regions, customs or farming systems. Since these costs *must* be met in any normal year, the main way to meet them is with the total gross margin earned from all farm activities, plus non-farm family income. (Of course, they could be met by selling assets, or going further into debt, but this can only be done for a short time and does not usually appeal to most farmers.) Total gross margins and non-farm family income should be equal to or more than total overhead costs.

As well as knowing the total value of the resources, it is useful to find out how much capital is tied up in each activity, e.g. the value of the chickens, cattle, goats, tools and machinery. This information can be used in planning which activities should be expanded, and which reduced. Another essential piece of information is the amount and type of debts, types of creditors and terms of the loans, e.g. short-, medium- or long-term. The amount of debt compared to equity, and whether the farmer could borrow more money if he wanted to invest in developing his farm are also useful bits of information.* Where it is relevant – such as in commercial farms, where land has a market value – a useful measure of the efficiency of the farm as a business is to express the annual profit it earns as a percentage of the total resources of the farm. This is known as the 'return on capital' (see Chapter 6 for details of 'profit').

There are numerous formal and informal arrangements the farmer, his family, and maybe the village can have concerning who runs what, how the food and 'profits' are shared, and who pays the taxes. It is necessary to find out about such arrangements. In the case of commercial farms, the structure (e.g. partnership, or company) has to be determined.

The financial potential of the farm under scrutiny can be evaluated after a number of issues have been examined. First, it may be possible to increase the gross margins of the existing activities by better 'management' (in the widest sense of the term), see Chapter 2. Second, it might be possible to change the mixture of

activities to produce more food or money, e.g. more cassava, less sweet potatoes. Third, there may be prospects for introducing *new* activities. If so, the expected profitability, the amount of extra capital needed, and the extra 'profit' has to be calculated. 'Profit' is, to some degree, what you think it is, and some definitions of profit are given in Chapter 6. Fourth, it may be feasible to rearrange loans with the lenders so that they are converted from short- to medium- or long-terms. This usually means lower annual repayments.

Fifth, where land has a market value it is worthwhile estimating whether there is likely to be an increase in its value each year after allowing for the effects of inflation. If so, this results in an increase in the farmer's wealth (even though he may not choose to sell his farm), and thus in his capacity to borrow funds to increase production. From estimates of the anticipated total gross margins and overheads, and of the future total value of the farmer's assets, the expected return to farm resources (see Chapters 4 and 6), can be calculated and compared with the present figures. Finally, there may be scope for changes in structure or arrangements for ownership, control and distribution of 'profits' which may lead to more harmonious and efficient working of the farm.

The main barriers to developing the full financial potential of the farm include lack of knowledge and management skills, the absence of any new technologies or activities, or the inability to utilise any new technologies or activities which will enable gross margins to be increased. Certainly, these are real barriers, but often they can be partially overcome with a better extension and research effort.

It may be difficult to obtain loans on more favourable terms, given the shortage of capital for lending in developing countries and the keen competition for it from the non-rural sectors of the economy. However, in many cases, farmers are able to get loans at rates which are cheaper than their business 'competitors' have to pay. There is also the problem that the staff at many lending institutions do not fully understand the cyclical nature of annual farm costs and receipts, nor the time taken for many farm development programs to become fully operational (often 6–8 years).

Tribal and family customs sometimes make it

*Equity is the amount of money a farmer would have if all the assets he owns were sold and he paid all of his debts.

difficult to change family arrangements or the 'business' structure of the farm, particularly the questions of who can use the land and who 'owns' it. Furthermore, there is a scarcity of extension workers who are trained in farm financial planning and in appraising a farm development proposal. The difficulty of getting sound and useful advice on the financial aspects of development works as a barrier to the farmer getting the most out of his farm.

Questions

1 Do you think we have covered all the main aspects of how to analyse and plan a farm?

2 If not, what other factors should have been mentioned?

3 What potential exists for improving the production, 'profitability', and asset worth of the farms that you know?

4 What steps are farmers able to take to exploit the potential of their farms?

5 What are some of the main restraints which prevent farmers achieving the full potential of their farms in your area?

4

Principles of production*

Introduction

When thinking about growing agricultural products it is necessary to consider what to work with, the most important needs, what can be expected to be produced in the area, and what can be done best or better than in other places.

If a person can work, then he or she has labour power which could be used as an input in farm production. Labour is a resource, or a 'factor of production'. Labour on its own does not grow food and fibre. In order to produce something of use, the worker needs other resources to work with (such as soil, tools and equipment, seeds, water, time, and some management skills). Capital is another resource. Capital has many meanings. For farm management purposes it can be thought of as physical goods (like tools and equipment), money, or the worth of what a person owns in his farm or other business.

Another broad class of resource is raw materials, which consists of the land and anything useful which it provides. Resources used in production (or 'factors of production') can be classed as land, labour and capital. None of these is plentiful when compared to the many purposes for which it can be used, and the many needs people have for it.

Technical principles

Production is the process of using resources to make goods, provide services, or to do both. Producers can use any or all of the three factors of production (labour, capital and raw materials) in different combinations, to produce one or many products. A key element in a farmer's decision about what to produce and how to do it is the objective of getting more, or

even the most, out of the limited amount of resources he has to work with. There are three basic production relationships (also called response relationships):

the relationship between the resource used and the amount of production (input–output);

the different ways resources can combine and substitute for one another in the production process (input–input);

the relationship between different products which can be produced (output–output).

(*Note*: output has the same meaning as product.)

Input–output

Let us consider the situation of a small farmer who has some land and a few tools (capital) to work the soil and weed the crop, using his own labour. Thus, he has land, labour and capital. He borrows some seeds, prepares a seed-bed and plants a crop. We will say that the soil has all the nutrients that the crop will need. He then needs enough rain and to weed and harvest the crop.

We are interested in the relation between the resources used (called input – in this case, labour, land and capital) and the production which results (called output – here, it is the amount of crop harvested). A

*The principles of production economics explained in this chapter derive from, and apply to, the general tendencies of farm 'firms' to behave in certain ways. This is how decision-makers in aggregate tend to act. The 'theory' does not necessarily give a prescription for each individual farm situation, because the data to apply it are often lacking. The 'extra cost, extra return' way of thinking is important for all decision-makers. For a further treatment of agricultural production economics, see:

(i) Upton, M. (1978). *Farm Management in Africa*, Oxford: Oxford University Press.

(ii) Doll, J. P. & Orazem, F. (1978). *Production Economics*. Ohio: Grid Inc.

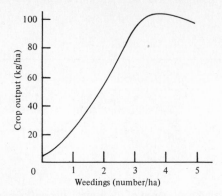

Fig. 4.1. Response function. The response function relates the number of weedings per hectare (applied over the life of the crop) to the total production of crop her hectare.

Number of weedings (per ha) (applied over the life of the crop)	Total crop production (kg/ha)
0	5
1	20
2	45
3	80
4	100
5	95

general principle of production is that where there is a fixed amount of one resource (e.g. land), then more output can be obtained only by adding other productive resources to it.

In this case land is the *fixed input*. It is called 'fixed' because once the crop is planted the farmer cannot, in the short term, change the amount of land being used. The land is used, regardless of whether a large or small yield is harvested. Other productive resource inputs such as labour, fertiliser or seed can be added to the production process in varying amounts. These are called *variable inputs*.

Let us look more closely at one of the things which affects crop yield – the number of weedings – in particular, the relation that exists between the total number of weedings done throughout the whole production process and the total amount of crop harvested. With all other influences held at some constant level, a direct link can be found between total weedings and total yield of crop. The relation between the number of weedings and the amount of yield shows the response of yield to weeding and is called a *response*

function (or a production function; see Fig. 4.1). This tells us that, for these particular conditions, if he weeds the land three times then the farmer may expect about 80 kg of crop. We are assuming here that each weeding is an input of similar quality and quantity. Unfortunately, actual response functions are not usually known with such precision.

A response function shows that the amount of output depends on the amount of variable inputs used with the fixed input, e.g. the amount of seed, fertiliser and labour, applied to the land. The land is going to be used for the crop, so the only factors to make decisions about are the variable inputs.

Let us look at weeding as though it were the only variable input to production. Without any weeding there would probably be very little crop production. The first weeding will cause quite a significant yield response. For the first three weedings, each extra weeding contributes more to total output than each previous weeding. This situation is called increasing returns. The extra production which comes from an extra unit of input we call 'the extra product'.* This is a key term.

At a low amount of the variable input (weeding), each extra weeding will contribute more to total output than each previous weeding. Two weedings may be more than twice as useful as one weeding. This is a situation of increasing returns, e.g. the first three weedings shown in Figs 4.1 and 4.2.

As weeding input increases, a level will be reached where each extra weeding adds less to total production than the previous weeding did. Each extra weeding still adds to total crop production, but each increase is less than the contribution of the previous weeding. Extra product is positive but getting smaller. This is called the *diminishing returns* stage of input use, e.g. weeding number 4, shown in Figs 4.1 and 4.2.

Ultimately, an extra weeding will not add anything to total crop production, but will cause it to fall. This could be due to damage to soil structure during the soil preparation stage, or damage by weeders to plants which are almost mature. At this level of input use, total crop production is suffering from too much of the variable input. The extra product resulting from each extra input is negative at this level of input use. This is

*Most economists use the word 'marginal' for 'extra', hence the terms 'marginal product', 'marginal cost' and 'marginal revenue'.

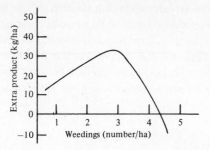

Fig. 4.2. Extra product. The extra product curve indicates the amount of extra crop arising from each extra weeding. The extra product is plotted at the mid-point between the number of weedings, as shown below.

Weedings (number/ha)	Total production (kg/ha)	Extra product (kg)
0	5	
1	20	15
2	45	25
3	80	35
4	100	20
5	95	− 5

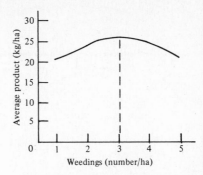

Fig. 4.3. Average product. The average product is the total product divided by the number of inputs (weedings).

Weedings (number/ha)	Total product (kg/ha)	Average product (kg/ha)
0	5	—
1	20	20
2	45	22
3	80	26.6
4	100	25
5	95	19

known as the stage of 'negative returns', e.g. the fifth weeding, as shown in Figs 4.1 and 4.2.

This technical relationship between variable inputs and outputs, where the extra output from extra inputs becomes less and less, is called the 'Law of diminishing returns'. This 'law' states: if increasing amounts of one input are added to a production process, while the use of all other factors is held constant, then the amount of output for each unit of increased input will eventually decrease.

Knowing details of this response of output to variable inputs, and without knowing anything about the cost of the variable input or the value of the output, it can be seen that some levels of input are sensible, and some levels of input are either too high or too low. For instance, it is not much good to weed the crop so often that the extra weedings reduce total crop production. In deciding how many weedings to do, a starting point is to ask: roughly, how many weedings will give the highest crop production from the land. If the land area which can be cropped is limited, labour is plentiful, and you need to grow as much crop as possible, then it is best to weed up to the level which gives the highest total

production from the land. In this example that will be four weedings. Up to four weedings the total output from the land increases with each extra weeding. Beyond this, total production decreases (Fig. 4.1). As well as looking at total product to decide how much weeding to do, it is a good idea to look at the amount of production resulting from *each* of the inputs used. This is called the average product (Fig. 4.3).

The average product of an input is the total amount of production divided by the quantity of the inputs that go into it. It is good sense to use inputs at least up to where the average product of the inputs (the weedings) is highest, i.e. three weedings.

Taking the example of weedings and land inputs to crop production, as more weeding (variable) input is applied to the land (fixed) input, total product increases rapidly. At first, extra weeding produces bigger and bigger increases in total output (extra product is increasing). When the extra product is increasing (adding more and more to the total produced), this is pulling up the average produce of all the weedings done previously, e.g. weedings 1, 2, 3 in Fig. 4.4.

The average product (AP) of an input is highest where the extra product (EP) equals it. Before this, EP

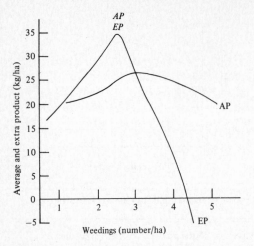

Fig. 4.4 The relationship between average product (AP) and extra product (EP) and number of weedings.

is greater than AP and pulling AP up. After this, EP is less than AP and extra weedings are pulling down the average production of all previous weedings. It is sound to use inputs up to where the average production from each weeding is highest. There is no sense doing less than this amount of weeding, because up to this point an extra weeding will increase the average product of *all* the previous weedings done. We have determined two levels of variable input use (weeding):

the amount of weeding which gives the highest *average* production for the weeding labour used;

the amount of weeding which gives most crop.

These define a range of input levels which are technically sound (zone 2, Fig. 4.5) regardless of costs and prices. As well, there are zones of production where too few (zone 1) and too many (zone 3) inputs are being used.

The economically 'best' level of input use will be somewhere in zone 2, but it also depends on the cost of doing the weeding, any cost of using the land, and the economic worth of the output.

It should be noted that the way we have shown the process of production in Figure 4.5 is quite a way removed from the reality of what happens in the field, the paddy, or the bush. Relations between the use of inputs and the resultant amount of outputs in any place

is never as precisely known as portrayed in Figure 4.5. Nor are the ranges of inputs which indicate 'good' and 'bad' zones of production as clear-cut as is shown.

Despite these drawbacks, the principles which underly the technical relations between inputs and outputs are valid. The view we have given of the

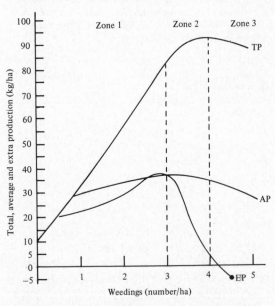

Fig. 4.5. Zones of production. Total product (TP), average product (AP and extra product (EP) define three zones of production. In zone 1, average product is increasing to a maximum and too few inputs are being used. In zone 2, extra product and average product are declining. The boundary of zone 2 is marked by total product reaching maximum and extra product reaching zero. The right amount of inputs is being used in zone 2. In zone 3, extra product is negative and total product is declining. Too many inputs are being used.

Weedings (number)	Total product	Average product	Extra product
0	5	—	15
1	20	20	25
2	45	22.5	35
3	80	26.6	20
4	100	25	−5
5	95	19	

production process should provide the basis for thinking about what to produce and how to do it. This approach to weighing decisions about using the things with which you have to work is a useful guide to making good farm management decisions.

Input–input combinations

How does all this translate to the situation where there is more than one variable input to use? Suppose that we have two variable inputs – weedings and quantity of seed planted. There are usually a number of ways in which these two inputs could be combined to produce a given amount of crop from the land. The farmer could combine a lot of seed with little weeding, or a lot of weeding with little seed, or use some combination between these extremes. Remembering the law of diminishing returns, combinations of a lot of both inputs could mean that the extra product of one or both inputs is negative. Such combinations would not appeal to most people.

The rate at which one input can *replace* another input in the production process, whilst maintaining a certain level of output, is of special interest. If the farmer does only one weeding, how much more seed needs applying to keep production levels up? If less seed is applied, how much extra weeding is needed? Inputs may *substitute* for each other at a constant rate (one for one), or more commonly, at a diminishing rate. We will start with one weeding and 200 kg of seed. One extra weeding applied to the production process might replace 90 kg of seed on a plot of land and still keep production up to 1 tonne; the third weeding might replace 40 kg of seed (Fig. 4.6). With diminishing returns, more and more of one input replaces less and less of the other input.

Different combinations of inputs which produce a certain amount of output can be shown as points along an *output line*. A 1 tonne output line shows all the combinations of the two inputs which combine to produce 1 tonne of output. This output line also shows the rate at which the two inputs substitute for each other (Fig. 4.6).

As in the single output case it is a good idea to use inputs in quantities where their contribution is dimin-

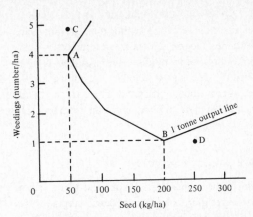

Fig. 4.6. Output from two inputs. Different combinations of weedings and seed inputs per hectare will combine to produce 1 tonne of output (represented by the 1 tonne output line). Combinations in the region AB are technically sound. Points C and D represent combinations with too much weeding and seed, respectively, causing output to drop below 1 tonne.

Input combinations[a]

Weedings	Seed (kg)
1	200
2	110
2	300
3	70
4	50
5	90

[a] Output: 1 tonne (1000 kg).

ishing but positive. In Fig. 4.6, this occurs over the range of the output line labelled AB. Point B represents the minimum amount of the weeding required to still produce 1 tonne of crop, combined with the maximum amount of seed which can be applied and still keep output at 1 tonne. With the minimum amount of weeding, any more than 200 kg of seed will result in less output, e.g. as in point D in Fig. 4.6. Beyond point B, with this amount of weeding, the extra production of extra seed is negative. Similarly, beyond point A, with the minimum possible seed (50 kg), more weeding would have negative returns and reduce output below 1 tonne (point C).

The input combinations covered by the AB portion

of the output line are technically sound. Here, in technical terms, seed or weeding time are not being wasted (similar to zone 2 in the single input case). These ideas hold for any number of resources used in production. The aim is to make good use of the farmer's land, labour and capital. These are scarce and can be used in many different ways, and should not be 'spent' unwisely.

Product–product combinations

The third case is where land, labour and capital can be used to make more than one product. Imagine a very simple case where it is possible to grow product X, or product Y, or some combination of the two. There are a number of ways that producing X might affect the production of Y (Fig. 4.7). The two products could help each other, meaning that if more X is produced it will contribute to more of Y being produced, i.e. X and Y are *complementary*. Or X and Y might be *supplementary*, meaning that more of one does not affect the amount of the other which is produced. Products X and Y could be *competitive* for the farmer's land, labour and capital (with more of one product, less of the other can be produced).

If a farmer wanted to make the best use of his resources then, on technical grounds, it would pay to make use of any complementary or supplementary effects between products, i.e. grow at least 10 X and 20 Y (Fig. 4.7). We have established that:

(i)   it is worthwhile to use an extra unit of input if it will raise the average output of the previous units of the input used; and

(ii) it is time to stop using more of the input if a bit more of it will reduce total product.

These principles 'work' and are useful ways of looking at production decisions for either fully subsistence or fully commercial farming. *But*, it needs stressing that in all these cases, single variable input, many variable inputs, and combinations of products, resources are unlikely to be as easily divided up into convenient little packages, nor as free to move from one use to another, as we have made out. The vital bit of information in all this, the *actual response function*, is unlikely to be known precisely. Furthermore, the expected levels of yields are not certain to occur. The best planned

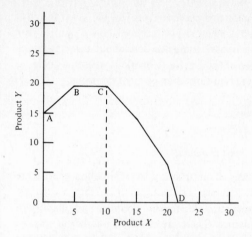

Fig. 4.7. Product Combinations. In the region AB the two products have a complementary relationship. In region BC the relationship between X and Y is supplementary. Within the region CD the two products are competitive at an increasing rate. The best combination of the two products lies somewhere within the competitive region.

Product Combinations

Product X	Product Y
0	15
5	20
10	20
15	15
20	5
22	0

production processes can be a total failure if expected rains do not arrive, or disease breaks out, or a plague of pests arrives; many other events can occur which will affect expected results.

What is the application of all this? Whilst some think that finding an exception or disproving part of a theory destroys it, we think that there are some technical truths of general relevance which can be put to practical effect. One such truth is that a 'cause-and-effect' understanding of the technical relations between amounts of inputs and amounts of output (and between different inputs and different products) provides a basis on which to make decisions about what to produce and how to do it.

How does one know these technical relations? There are two important sources – the farmer and the researcher. Farmers have a reasonably good idea of the size of the effects which different amounts of input will have on output. Researchers can know the relationship between inputs and outputs under experimental conditions. Their results, if interpreted cautiously to allow for the differences between experimental conditions and the real farming situation, can help farmers and extension workers.

These technical relations of production are the starting point of farm management decisions. By knowing the output the farmer might get by using certain inputs, plans can be made which indicate to the farmer what would happen if the expected outcome does occur (and if it does not). With this sort of information economic analyses can be performed.

Economic principles

Some general points

A starting point in economics is that there is not much sense in doing something if the end result is that you end up being, in some way that you regard as important, worse off than you were before you started. Something is usually worth doing if you strongly believe that you will end up better off, with aspects of your life changed in some way that is worthwhile to you.

In the economic approach, emphasis is on making the most of whatever is desired out of the limited resources available. A key idea is that the producer uses resources in such a way as to have a 'bit more' of something, to be better off in some way. The 'something' focussed on is often money (income) because it is easily measurable, but the principles can equally apply to quantities of production, or peace of mind for the farmer because risk is less, or even farmer 'satisfaction', which could encompass all of these measures. Satisfaction cannot really be measured.

Getting the most output from limited land, labour, and capital will not be the only objective, or even the major objective of farmers. Farm production decisions may be made with the first aim of guaranteeing sufficient food production from each production period. A cropping regime may have to: (i) supply enough food of the various types required, (ii) reduce the risk of not supplying enough food and, if the first two needs are met, then (iii) to get most 'profit'.

Part of the study of economics concerns using resources efficiently. Efficiency refers to a *ratio* of what is used to what is produced, e.g. the amount of corn which needs to be fed to an animal in order to add to it 1 kg of weight. Making the most of limited resources means getting the most output from inputs (technically efficient) and achieving as many financial and personal objectives (economically efficient) as possible.

Economic principles can be used in deciding what to produce and how to do it. That is, how to choose between alternatives, how to allocate limited resources among alternative uses. As soon as *choice* is involved, economics is relevant. If you choose to use resources in a certain way then you have given up the opportunity to use your resources in some alternative way. Making choices means opportunities are given up, opportunity costs are involved. The concept of opportunity cost is useful in helping to decide 'which is best?' If I do this, what am I giving up the chance to do? Which choice makes me better off?

Suppose that there are alternative uses for the resources available to a farmer. The land could be used for different crops, or for animals, or sold or leased to someone. He could sell his tools and use the money to build a room on the house, or to lend, or to buy different tools. He can exchange the produce he grows in a market for some other commodity, he could work elsewhere and hire someone to do the weeding and seeding. His land, labour and capital all have alternative uses and are of different worth in different uses. He must pay to get control of those resources. The situation we are describing includes the possibility of buying and selling some inputs, and some output, in markets. *In theory*, in competitive markets there are lots of buyers and sellers all with good knowledge of what is happening in the market. The individual producer's produce is not any different from anyone else's, and there is no individual buyer or seller big enough to influence the bargaining and trading processes.*

If this is the case, and it is more likely in farming than other industries, then the markets are probably

*This may not be true, in reality, with respect to the suppliers of some of the farmer's purchased inputs (in particular fertilisers and chemicals) nor if there are only a few buyers of the farmer's production, such as a couple of not very competitive fruit-processing firms. Western economic theory tackles those situations with theories about *imperfect competition*.

fairly competitive, and the price formed for the inputs and the output sold is a reasonable guide to how much producers and consumers want to have command over different resources, and products, for different uses.

Western economic analysis has developed the idea that if this competitive market happens then certain economic principles, with consequences for resource use and efficiency, will follow. These principles are applied to agricultural production in the rest of this text, and they 'work' to the extent that competitive conditions of production apply.

Producers may need some way of comparing alternative uses of resources. Comparisons require common units of measurement. In economics, money is usually used as a measuring stick. So, prices have to be taken into account when making choices about what to produce and how to produce it. The costs of inputs and prices of outputs are used to calculate how much production, and what combinations of inputs and outputs, are best.

The technical principles of diminishing returns and marginal effects in production outlined so far in this text are at the heart of economic analysis. For economic analysis it is necessary to relate technical knowledge of production to the costs and returns associated with production. Economic ways of thinking, or economic ways of looking at and seeing what is happening, analyse problems and make decisions on the basis of causes and expected effects.

> If I apply extra fertiliser which costs a certain amount and it increases total production, is it worth doing? or
>
> If I use this land to grow this crop instead of another crop, in what ways will I be better or worse off?

These are economic ways of looking at farm production choices and deciding what to do. Of particular interest are changes to inputs and outputs 'at the margin' – the idea of a little bit more, a little bit less. Farm management decision-making is, in part, the skill of applying production principles to the use of physical, human and financial resources in producing agricultural goods. It is about tackling the problem of 'the resources are scarce, our needs are many and varied, and we have to make the best use we can of our

resources', as it applies to farming decisions and farmer's objectives.

A key connection to make is the idea that, since economic and technical principles underlie a lot of the things which happen in farm activities, they *also* underlie the budgeting and planning tools used in decision making, described in later chapters.

Deciding on resource use

One important question is, 'how can information about response functions, costs and prices be used?' It can be used to work out whether a farming activity, or a change, is going to make the producer better off in some way. The 'better off' we are talking about in this case is 'profit'. By 'profit' we mean some notion of a surplus of the output of proceeds from production over all of the costs incurred in carrying out that productive activity. It is useful to see how to make a profit (and even make the maximum profit possible) using limited resources.

Consider, again, the case of the farm family with some land, labour and capital with which to grow a crop. In the following discussion it is assumed that the resources used in production have a cost, that output can be sold for a price, and that the farmer aims to make as much profit as possible. A feature of economic ways of looking at a situation is thinking in terms of a little bit less of this, a little bit more of that, to improve profits. We demonstrate this approach in the rest of this chapter.

Input–output

Where there was only one variable input (weeding labour) we showed that the extra output from extra weeding at first increases, then declines. Further, the best level of weedings lies somewhere in zone 2 (see Fig. 4.5). Knowing the cost of each extra weeding, knowing how much extra output will be produced by each extra weeding, *and* knowing what that output is worth (how much it could sell for or what it would cost to buy it), it is possible to work out which level of weeding will make most profit. With more weeding, where extra outputs are still positive but becoming smaller, the cost of extra weeding may eventually exceeed the value of the extra output which results. If so, further weeding is

Table 4.1. *Extra cost and extra return*

Weeding number	Cost of each extra weeding ($)	Output from each extra weeding (kg/ha)	Return from each extra weeding ($)	'Profit' from each extra weeding ($)
1	10.00	100	50.00	40.00
2	10.00	150	75.00	65.00
3	10.00	120	60.00	50.00
4	10.00	50	25.00	15.00
5	10.00	25	12.50	2.50
6	10.00	10	5.00	− 5.00

not worth doing. It would reduce the total profits which have been made from all the previous weedings (see Table 4.1).

Weedings 1, 2, 3, 4, 5 are each worth doing as each adds to the total profit. Weeding 6 costs more than it brings in, reduces the total profit, and is not worth doing. A decision rule is to weed the plot until the extra cost of weeding just about equals the extra return from doing so, i.e. weeding No. 5 where extra revenue = $12.50, extra cost = $10. Too few weedings will mean you do not get some profits which could be made, and too many weedings reduces the profits you could have made. If you can afford it, the rule is to use inputs to where their extra return equals the extra cost of using them.

Theoretical best levels of inputs and outputs are guides to resource use, showing the way in which the use of resources ought to be moving. What is expected to happen might not occur, and the farmer has to be prepared to reduce the bad effects if 'the worst' happens. The notion that extra return should equal extra cost points to the direction of wise input use, rather than being a practical 'best'. If the farmer cannot afford this best level of inputs and if there are no other uses for the inputs and no other crops to grow, then the decision rule is to use as much input (in this case do as many weedings) as can be afforded.

Input–input combinations

Let us look at the situation where there is more than one variable input (e.g. weeding hours and seed). The most economic combination of inputs depends on the cost of each input and the relationships between them.

The aim is to combine these two inputs so as to produce output as cheaply as possible. From the costs of the two inputs, the total cost of different combinations of the inputs can be calculated (Table 4.2), and the cheapest combination(s) found. If one weeding costs $10 and seed costs $0.50/kg then 3 weedings and 70 kg of seed, or 4 weedings and 50 kg of seed each costs $65.

When the price of an input changes, so does the cheapest combination. If the price of weeding labour rises relative to the price of seed, then the farmer would try to produce the same output using less weeding labour and a bit more seed.

If a lot of one input is being used and its extra contribution to output is becoming 'low' (remember diminishing returns), and little of another input is being used and its extra return is still 'high', then in order to produce the same output more cheaply, use less of the 'low' extra-return input and more of the 'high' extra-return input.

To make best use of limited funds, for any particular amount of output, swap and substitute the inputs until the cost saved by using less of one input is about the same as the extra cost of using more of another input. Here we needed to make a particular amount of output.

Product combinations

Finally, what is the profit-making rule when producing two or more products? If the products are complementary or supplementary then it is good technical sense to

Table 4.2. *Input combinations*[a]

Input		
Weedings	Seed (kg)	Total cost ($)
1	200	110
2	110	75
2	300	170
3	70	65
4	50	65
5	90	95

[a] Output in each case equals 1000 kg.

Table 4.3. *Combinations of products and total revenue*

Number of product X	Number of product Y	Total Revenue (Worth of $X=\$4$, $Y=\$5$)
0	15	75
5	20	120
10	20	140
15	15	135
20	5	105
25	0	100

produce both products at least up to where they start to compete for the limited resources.

A farmer may be able to produce either all product X, or all product Y, or some combination of the two with the bundle of resources he has to use. As shown in Fig. 4.7, these two products are competitive within the range AB. Here, more X can be produced only by producing less Y. Suppose that product X returns \$4 and product Y returns \$5. Different combinations of X and Y will bring in different amounts of money, yet if they are produced with the same resources, they cost the same to produce. The best combination then is the one which brings in the most money. To find the combination which brings in the most money, look at the technically possible combinations and the price of each product. Then calculate the total revenue each combination will bring in (Table 4.3).

Production of 10 of X and 20 Y will bring in most revenue. Changes in the relative prices of the two products will change the best combination.

At the combination of two products which brings in the most money, the extra return from using available resources to produce more of one product is about the same as the extra return from using the resources to produce more of the alternative product. Here, there is no scope for further gains by substituting more of one product for less of another. This is an application of the principle of equi-marginal returns.

It may be that by the time all of the technical requirements for the production of several complementary crops are met, there is little scope to make more money by adjusting the combination of crops grown and resources used.

Example The following example illustrates the principles of making most 'profit' when funds are limited and

there are alternative uses for them. Assume that the farmer grows three crops – 1, 2 and 3 – and that prices are as follows:

Crop 1   20¢/kg
Crop 2   16¢/kg
Crop 3   8¢/kg

Fertiliser costs 24¢/kg.

Yields and financial returns for a range of fertiliser applications are shown for each crop in Table 4.4. The fertiliser is applied in 100 kg lots, each costing \$24.00. The bold numbers circled show the last stage where extra returns exceed extra costs.

All of this information is often difficult to obtain in practice, but the example is included to help illustrate a basic principle of resource allocation.

Were there plenty of funds, the most profitable amount of fertiliser to use is where the extra return just exceeds the extra cost of the fertiliser, i.e.:

Crop	Total amount of fertiliser applied	Extra cost of each 100 kg of fertiliser applied (\$)	Extra return from the last 100 kg of fertiliser applied (\$)
1	600	24	28
2	300	24	28.80
3	700	24	25.60

In Table 4.4, we have tried to show how best to spend the scarce dollars available for fertiliser. There are three competing uses for these scarce dollars, viz. crops 1, 2 and 3. The supposed yield response of the three crops to the fertiliser is shown on the left hand side of each *major* column of the table. For example, under 'Crop 1', if you use 100 kg of fertiliser, this (hopefully) would give a yield of 1300 kg of that crop, 200 kg of fertiliser should produce 1640 kg of crop and so on. You keep applying fertiliser until the extra return *just slightly* exceeds the extra cost (see the numbers set in bold). This allocation of fertiliser, costing \$384.00, would return \$1308.

More likely, funds would not be available to apply this much fertiliser to each crop. What is wanted is to make the most use of the limited funds which are available for spending on fertiliser. This is done by first

Table 4.4. *Additional yield and return from a range of fertiliser applications to 3 crops*

Fertiliser			Crop 1			Crop 2			Crop 3		
Application per ha (kg)	Cost per ha @ 24¢ ($)	Extra cost per ha ($)	Yield per ha (kg)	Return per ha ($)	Value of extra product per ha ($)	Yield per ha (kg)	Return per ha ($)	Value of extra product per ha ($)	Yield per ha (kg)	Return per ha ($)	Value of extra product per ha ($)
0	—	—	900	180.00		1000	160.00		2000	160.00	
100	24.00	24	1300	260.00	80	1330	212.80	52.80	3040	243.20	83.20
200	48.00	24	1640	328.00	68	1580	252.80	40	3940	315.20	72.00
300	72.00	24	1920	384.00	56	1760	281.60	**28.80**	4710	376.80	61.60
400	96.00	24	2150	430.00	46	1890	302.40	20.80	5360	428.80	52.00
500	120.00	24	2330	466.00	36	1975	316.00	13.60	5900	472.00	43.20
600	144.00	24	2470	494.00	**28**	2020	323.20	7.20	6330	506.40	34.40
700	168.00	24	2570	514.00	20	2020	323.20	0	6650	532.00	**25.60**
800	192.00	24	2638	527.60	13.60	1985	317.60	−5.60	6870	549.60	17.60
900	216.00	24	2676	535.20	7.60	1935	309.60	−8.00	6770	541.60	−8.00

Table 4.5. *Best allocation of fertiliser to the 3 crops*

Cumulative fertiliser cost ($)	Extra return from each $24 of fertiliser applied (in 100 kg lots/ha)		
	Crop 1 ($)	Crop 2 ($)	Crop 3 ($)
24			83.20 (1)
48	80.00 (2)		
72			72.00 (3)
96	68.00 (4)		
120			61.60 (5)
144	56.00 (6)		
168		52.80 (7)	
192			52.00 (8)
216	46.40 (9)		
240			43.20 (10)
264		40.00 (11)	
288	36.00 (12)		

applying fertiliser to the crop where the extra return is highest, and then to the crop which has the next highest return, and so on.

Suppose the farmer had only about $300 available for fertilising three possible crops, each to cover 1 ha. The aim is to make the highest profit from the limited amount of money he has to spend on fertiliser. He has two basic choices (i) to apply 4 (100 kg) units of fertiliser to each crop (the 12 units cost together $12 \times \$24 = \$288$) and (ii) to apply the fertiliser on the lines indicated in Table 4.5. The principle here is that the first unit of fertiliser should go to the crop which gives the highest extra return, the next unit to the crop which gives the next highest return, and so on. Note that the data in Table 4.5 are derived from those in Table 4.4. We have shown, by the numbers in brackets in Table 4.5, the priorities which should be given to each (competing) crop. Thus, the number '1' beside the $83.20, under the column for crop 3, on the row for the $24 fertiliser cost, indicates that the highest return for the first $24 unit of fertiliser will come from applying it to crop 3. Where should the next unit of fertiliser go? To crop 1, because it gives an extra return of $80. That is why there is a number '2' beside the $80. The next unit goes to crop 3, the next to crop 1 and so on. Were the $288 of fertiliser divided equally between the 3 crops, the total income would be $1161. By following the principle of allocating it in the way we have

suggested, the income is $1190. Though the difference in this example may appear trivial, the *principle* is not. It has application in any situation when funds are scarce and there are competing uses for them. For example, a government has to decide how its limited money should best be shared between say farm development, roads, expanding electricity and subsidising food for urban consumers.

Time

Production takes place over a period of time – time during which many things can happen to disrupt the planned production, time during which resources could be used for something else for a different reward, time during which technical changes can alter production relationships. As well, a dollar received or spent today is worth more than a dollar some time in the future because (i) today's dollar can be used to earn income or satisfy some needs and wants, and (ii) inflation might reduce the amount of goods and services that a dollar can buy in the future. This means that the rules for making good use of limited resources still apply, but extra costs and returns have to be adjusted for the effects of time.

Relevance of economics to tropical farming

So far we have discussed:

production relationships which define ranges of input uses and combinations of output which make good sense in a technical way;

economic aspects of production relationships which define the most efficient and most profitable levels of input and combinations of output.

Unless there are some realistic and reasonably significant opportunities for using some or all of these resources in some alternative way, or for trading a significant proportion of inputs (e.g. labour) and outputs in markets, then economics does not come into the production decisions.

Production economics is about using resources in the 'best way'. The phrase 'best way' implies that there is more than one use for the resource, and that significant choices exist. In a subsistence farming situation where the farmer has an area of land which he has a right to use for farming,

he has some simple home-made tools to use for cultivation, weeding and harvesting,

if he does not use the land to grow a particular crop, then there is nothing else the land could be used for,

there are no opportunities to work elsewhere,

if he does not dig up the land, plant and tend a crop, he will do nothing at all,

the crop that is harvested will be stored for consumption through the next cropping season,

then there are really very few important choices to be made about the 'how' and 'what' aspects of resource use and production. The 'when' aspects of production are also important to decisions; in many environments there is little choice about 'when to do what'. Under the subsistence system described above the farmer faces a very limited number of choices, and a limited range over which the choices apply – mainly a couple of technical choices between the amount of seed per hectare, the amount of weeding to do, and the best time at which to do various operations.

The choice between the amount of seed and the amount of weeding depends on the known or best guess about technical input–output responses and the farmer's needs and inclinations. If his objective is to be fairly sure of growing enough food for the next season, without having to give up too much leisure, or social obligations and activities, and if weeding labour is in short supply, and seed plentiful, then he might go for a higher seeding rate and a bit less weeding. Whichever way he decided to produce the food he needs, it does not really matter in terms of resource use because there is very little else his land, labour, time, managerial skills, equipment and seed could be used for. Within reason, any way which ensures the required food will be produced and enables the farmer to achieve other important aims, is the 'best' way.

So far, we have used the extreme example of a purely subsistence farmer with no real production alternatives or major resource use choices. In reality, few farmers are *strictly* in this position, though many farmers might not be in a very different situation. Most farmers can be classed as semi-subsistence; they produce to meet their own basic needs, and also have

an eye on the chance of selling some products to be able to buy some other products. To decide if economics has much to say about the small farmer's situation, first look at the situation in which production economics has a lot to say.

Production economics is a powerful analytical tool in situations where there are many alternative uses for resources which are relatively scarce; where inputs and outputs are bought and sold in markets made up of large numbers of competing buyers and sellers; where individual products traded are not better or worse than other products of the same kind; and where all buyers and sellers in the market are well informed.

The less the extent to which these conditions of competition and choice occur in the small farmer's situation, the less useful are production economic analytical tools and decision making techniques. But even here, they are never useless. The economic basis of management decision-making is only as appropriate as the degree to which a decision maker's situation resembles the particular circumstances in which production economic analysis is likely to provide a guide to action.

Key points to consider are whether (i) small farmers face significant choices as to how they use their resources and what they use them for; (ii) have incentives to make best use of the available resources; (iii) face significant production uncertainty and prefer more certainty than less; (iv) have the option of selling or trading a significant proportion of their outputs in markets; (v) have the option of or need to purchase important farm inputs in markets; (vi) have a strong need or want to be better off; (vii) need to achieve or maintain output levels despite severe shortages of some major resources; (viii) have new production technologies becoming available to some of them; (ix) have access to sources of credit; and (x) have a social and economic system of organisation which involves substantial use of markets to bring together producers and consumers.

Our conclusion is that there are enough answers of 'some probably do, some probably will' to outweigh the 'most don't, most never will' answers to these questions. There is enough, we think, of the theory of production economics which is of relevance to *some* farmers/advisers in *some* situations to be of *some* use. This is why we have tried to explain techniques of farm management decision making which are based on the principles of production economics. In the following chapters we show how, with modification, these principles and decision techniques can be of some use in advising the small farmer.

Questions

1 Give some examples of the application of the principle of diminishing returns on farms that you know.

2 Has the principle had any practical application?

3 What are the main factors limiting its application on those farms?

4 'The stage of maximum physical production (output) is rarely the most profitable stage in any enterprise, although in practice, the two stages are often very close'. Show your understanding of this statement by illustrating how you would calculate the point of maximum profit for any given activity.

5 What do the terms 'input–input', 'input–output' and 'output–output' mean? Would you find them useful if you were advising a semi-subsistence farmer on his crop or animal production practices? If not, why not?

6 In the example given on pp. 34–6 we showed how best to spend scarce funds when there were three alternative uses for them. Can you think of other applications for the principle illustrated? What are they?

7 What is a 'resource'?

8 Do you think, from reading pages 36–8 that we are correct in concluding that ideas of production economics are relevant to the situation of the small farmer in the tropics?

5

Costs and returns*

Costs in theory

We will first make some general, 'theoretical' points
about costs. It is hard to say precisely just what all the
costs of owning and operating a farm are. To some
extent, costs are what you think they are. Like all other
farm figures, some costs are obvious and easy to
measure (such as cash out-of-pocket, expenses) whilst
others are 'hidden' and can only be roughly guessed.

Hidden costs include non-cash costs like the cost
involved if a farmer owns a piece of machinery which
will last a number of years, but is slowly wearing out. It
wears out a bit at a time with each crop it is used to help
produce. So, with each crop there is a 'wearing-out
cost' of capital equipment. It is called depreciation.
The annual cost of owning and using a machine will
depend in part on the length of its working life. A
judgment (guess) has to be made about the length of
working life which the machine is likely to have (see
Chapter 13). 'Opportunity cost', which is the revenue
which could have been earned but is given up when a
decision is made to use resources in an alternative way,
is another hidden cost. So, whilst some costs are
matters of fact, others depend on opinions and
assumptions.

Fixed (overhead), variable and total costs

Actual costs of production are the sum of two
components: (i) fixed (or overhead) costs which are not
directly related to the amount of crop produced on the
land resources (they have to be paid whether anything
is produced or not and include land rent, land taxes,
loan repayments, living expenses); (ii) variable costs,
which are directly related to the amount of crop grown
and so with the amount of variable inputs used (e.g.

weeding, labour, seed, fertiliser). Adding total fixed
costs and total variable costs gives the total cost of
production.

As fixed costs have to be 'paid' whether production
occurs or not, the fixed cost component of the total
cost of producing 1 unit of output will be high
compared to the fixed cost part of the total cost of
producing 10 units of output. The more output is
produced, the lower the amount of fixed cost involved
in the making of each unit of output (fixed costs are
spread over more output). So, average fixed cost tends
to fall as the quantity of output increases (see Fig. 5.1).

Variable costs are the costs of the variable inputs
used. As more and more variable inputs (e.g. fertiliser)
are used, then each extra input adds less to output. It
takes more and more input to produce an extra unit of
output (principle of diminishing returns). This means it
eventually requires more and more variable costs to
produce an extra unit of output, i.e. average variable
cost tends to increase as output increases (e.g. fertiliser
per kilogram of grain). Similarly, the total cost of each
unit of output (called average total cost) is made up of
average fixed cost plus average variable cost. In theory,
falling average fixed costs and rising variable costs
combine to create a U-shaped average total cost curve
(see Fig. 5.2).

Average cost of production highlights the gains
which can be made from having a size of farm
operation which is large enough to spread the fixed
costs, and produce each unit of output more cheaply
than is possible with a smaller sized business. This is
called achieving economies of size, where the average
total costs of production are reduced (Fig. 5.3). In

*Sometimes the words 'receipts' and 'returns' are used interchange-
ably. 'Returns' is used here for cash and non-cash items, 'receipts'
simply for just cash.

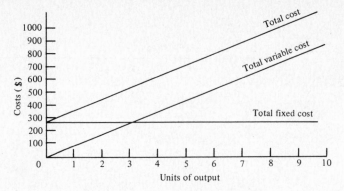

Fig. 5.1. Cost lines: fixed costs are constant regardless of output; variable costs vary directly with the amount of output. The total cost is the sum of the total fixed cost plus the total variable cost.

addition to the inefficiencies arising from being too small, there are also possible inefficiencies from being too large, i.e. when average total costs start to increase (diseconomies); these diseconomies could occur if problems of management arise as size increases (Fig. 5.3).

Another application of average fixed and variable costs of production is in decisions about continuing or stopping production in times of temporary adversity. Fixed costs have to be met regardless of whether anything is produced or not. A farmer has more decision-making choices over variable costs, in the short run. If things were going badly and he was not making enough money to pay all his costs, then he would have to think about whether or not to continue producing. If he is more than covering variable costs, so earning something towards meeting the fixed costs and hence reducing the final loss, then he would be better off if producing than if he were not producing (at least for a short while, see Fig. 5.4). If he were not even covering variable costs, then he would be better off not producing. Continuing to produce further adds to his unavoidable loss of fixed costs (i.e. some variable costs are lost as well).

Opportunity cost

The most hidden (i.e. most difficult to quantify) cost might be the type we call an opportunity cost. This is a cost which is incurred because money is used to provide equipment, seed, hired labour, in order to grow a crop.

Other opportunities for using these funds are thus given up. For example, this money could have been used to grow a different crop, or it could have been lent to someone who would pay interest on the loan. The opportunity cost of putting funds into one use instead of another must be considered.

The following procedure is recommended: once a budget of the actual expected costs and returns is done (not including opportunity costs), the outcome of the budget based on actual costs and returns should be examined in the light of some realistic, possible other uses of the resources (i.e. the opportunity costs), if any exist.

The same thinking applies to a farmer's own labour. If (i) there were a real chance for him to do something else with his own labour other than growing and harvesting a crop, and (ii) it were likely that he would consider doing something else, then the opportunity cost of using his labour in one way rather than another, ought to be taken into account.

There is often an opportunity cost not only for resource use between two different activities, but also within a single activity. For example, assume that a farmer could grow 100 kg of crop if he did five weedings and only 75 kg if he did four weedings. If he decided not to do the fifth weeding and instead to have a bit more rest and play more cards, then the cost of resting or playing cards is the 25 kg of crop foregone by not weeding. An opportunity cost, if it is real, and if it is known, can be a useful measuring stick for comparison when deciding if something is worth doing.

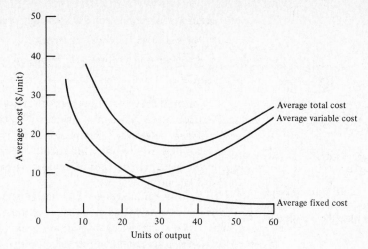

Fig. 5.2. Average fixed, variable and total costs.

Costs in practice

Apart from the value of cash flow figures, it is also important (especially for planning purposes) that a commercial farmer or farm adviser understand the nature of all the different types of costs and receipts in a farm business. Not all costs and income are of the same type. For instance, $50 spent on clothing has no effect on farm profit, but $50 spent on fertiliser has. Similarly, income from sales of a crop is of a different nature from income received from the sale of a machine that is no longer needed. It is helpful to understand such differences because they are important when planning future farm activities. A brief explanation of each main type of cost and receipt will be given here. The six main kinds of farm costs are: variable, overhead, finance, capital, personal and tax.

Variable costs

Variable costs are also known as direct costs. As the name implies, they are costs which vary as the size and/or level of output of an activity varies. For example, if the area under maize is increased by 50% then seed, fertiliser, and labour inputs will also increase (though not necessarily by about 50%). Sometimes, for example, labour may be available without extra charge. If cattle numbers are doubled, variable costs such as feed, and medicines will also double (approximately). Typical examples of variable costs are:

fertiliser;
seed;
sprays;
bought feed;
animal replacements;
harvesting;

cost of seasonal labour;
fuel and oil;
bags;
repairs to machinery and plant;
running irrigation plant.

Variable costs are proportional to the level of intensity of each activity, but may also determine the yield (level of output) of that activity. Thus (with crops) the amount and kind of fertiliser, seed, cultivation and weeding largely control the crop yield. Similarly (with animal activities) the level and type of feed, and the type of 'medicines' used have a major effect on the productivity of any given type of animal.

Other types of variable costs do not determine yield or production, but are necessary for the harvesting, transportation and marketing of a crop. Very little output occurs on farms unless money is spent on variable cost items. Conversely, even though a large amount of money is spent on overhead costs (defined below), most of this has little effect on the level of crop yield or animal production, because overhead costs are not related to a specific activity.

The reason for identifying the variable costs of an activity is to give the farmer an idea of how much costs will change if he expands or contracts any activities. For example, if he decides to decrease the cultivation area of cotton and to increase the area of maize, the variable costs will change but the overhead costs are

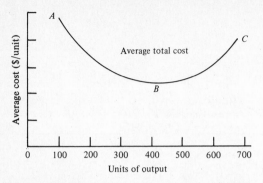

Fig. 5.3. Average total cost. Economies of size are being realised when the average total cost of output is declining (*AB*). Diseconomies of size occur when the average total cost of output is increasing (*BC*).

likely to remain about the same. Knowing the likely variable costs and gross income, the farmer's adviser is in a position to make a quick assessment of the merit of any proposed change in activities.

Overhead costs

Sometimes known as 'fixed' costs, overhead costs are those which, within limits, do not change when the level of activity changes. Thus an increase of 20% in the area of a crop, or in the number of animals, is not likely to lead to a rise in overhead costs. An increase as great as 100% would, however, increase overheads. On most farms, the overhead costs do not change very much as the level or mixture of activities changes except, of course, for increases due to rising costs. Two types of overheads are generally recognised: total and operating overheads.

Total overhead costs

Total overhead costs include:

essential living expenses of the farmer;

wages, food and clothes for permanent workers and their families;

loan interest and repayments;

replacement of capital items such as tools, machinery, buildings;

all taxes;

repairs to water supply, roads and structures;

insurances on employees, fixed structures and plant;

travel and other 'business' expenses;

costs of running a motorcycle or pickup;

rent.

Sometimes the wages of any permanent workers are allocated to the different activities and treated as activity variable costs. This procedure is usually a waste of time as it provides little additional information for planning purposes. This does not apply to casual labour, of course.

The main advantages in the farmer and his adviser knowing the level of total overheads is that they are the unavoidable costs which must be met each year. This makes clear the minimum total gross margin* which must be achieved from all the farm activities being considered. If the expected total gross margin from the present type and intensity of activities is not enough to cover the total overhead costs, the farmer should modify his plans to produce a programme which will cover them. Otherwise his debts will increase, perhaps to a dangerous level.

The total gross margin is normally the only source, other than additional borrowing, from which the overhead costs can be met. The total gross margin can usually be increased only by (i) wise spending on more variable-cost-inputs such as fertiliser, and (ii) using improved methods.

Operating overheads

Operating overheads are used in calculating the 'true profit' in a financial accounting sense. They are the overheads associated with the annual business operation of the farm. For accounting purposes, we do not include repayments of loans, interest, living expenses, or income tax in operating overheads. However an 'operator's allowance' for the work done by the operator is included. Also considered is the decline in value (or depreciation) of all capital items such as tractors, rather than the actual cost of replacement. The main components of operating overheads are itemised here:

*Total gross margin is total gross income minus total variable costs. This is explained fully in Chapter 7.

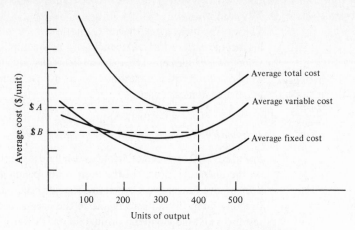

Fig. 5.4. Reducing total losses. If not earning \$A per unit of output, but earning more than \$B, then variable costs are being more than covered. The size of the total loss is lessened by producing, for a short time at least.

'operator's allowance' – a controversial, often irrelevant concept for semi-subsistence farming;

depreciation of capital items such as buildings and machines;

wages of permanent workers;

taxes other than income tax;

repairs to water supply, roads, buildings and structures;

insurance for employees, fixed structures, plant and buildings (if available);

business expenses;

costs of licensing and running motorcycle, pickup.

Note: Operating overheads + total variable costs = total operating costs.

Finance costs

Finance costs cover the annual interest paid on borrowed money, and the repayments made on loans. Where hire purchase is used, the payments include interest, loan repayment and often some insurance costs lumped together in one sum.

Capital costs

Typical items of capital expenditure are new buildings, tools, machinery, land purchase, land clearing, water supply, extra livestock and planting of palm oil, rubber, cocoa or fruit trees. Funds spent on capital

items should, but do not always, increase the productive potential and asset value of the farm, i.e. \$1000 spent on a capital item may add only \$700 to the market value of the farm.

Most capital items lose value, or depreciate, over time, and a depreciation allowance should be deducted from gross income each year so that the item can be replaced at the end of its useful life. The simplest way of calculating depreciation is to use the 'straight line' method, which assumes that an item loses value by the same amount each year. However, in many countries the cost of many working capital items such as breeding stock, irrigation equipment, and farm machinery tends to increase both in money and in real terms as time passes. So it is best, from a management and planning point of view, to use expected replacement cost, rather than purchase cost, as the amount which should be depreciated. As a simplified example:

	Cost (\$)
Purchase price	2100
Expected net replacement price	2800
Expected life: 7 years	
Annual depreciation $\dfrac{\$2800}{7} =$	400

(Not \$300, which would be the figure if purchase price was used.)

(See Chapter 13 for the full treatment of this issue.)

Table 5.1. *The contribution of variable costs to net profit*

Operating overhead cost ($) 1	Variable cost ($) 2	Total operating cost ($) 3 (1 + 2)	Gross income ($) 4	Operating profit ($) 5 (4 − 3)
6	1	7	3.00	− 4.00
6	2	8	5.90	− 2.10
6	3	9	8.70	− 0.30
6	4	10	11.20	+ 1.20
6	5	11	13.40	+ 2.40

Where there is a resale market the salvage of sale value of the piece of equipment can be taken into account. It is important that the smaller farmer understands the need to make allowance for replacement of capital resources such as work animals or small plant. Otherwise, when their working life is over he may not be able to find the money needed. It is the job of his adviser/extension worker to make him aware of this in his planning. Normally, replacement funds are used in the farm business because of pressing needs or because the return from using them is higher than they could earn in a bank, for example. However, such funds should be easily available either directly, or as items acceptable as a security for borrowing when the time of replacement falls due.

Household (personal) costs

Purchased food, clothing, medical expenses, school fees and family travelling costs are considered as personal costs. In certain cases, some of them are directly related to the level of output of the farm. For example, an ill or undernourished farmer and his family are not likely to have a high work output, and money spent on food or medicines is likely to have a direct effect on total farm output. However, in many other instances (e.g. schooling and clothing), such an effect is hard to measure. The minimum total living or personal costs of the farmer are normally included in the total overhead when budgeting for family farms, as they are one of the most important and unavoidable items in total farm costs.

Tax costs

The way in which taxes are levied on the farmer and/or his income varies widely. Depending on how complicated the arrangements are, they can be treated either as an overhead cost or as separate classes of cost, such as cash or non-cash costs.

Relation between costs, income and operating profit

Spending $200 on a capital item has a different effect on the operating 'profit' of the farm from spending $200 on overhead costs. Capital investments occur at irregular intervals and usually add to the productive resources and the long-term profitability of the farm. Overheads are annual, largely unavoidable costs which do not contribute to income in the short term, though they are an essential prerequisite to the ability to earn income, e.g. non-payment of taxes could lead to forced removal of the farmer from his land. Similarly, $200 spent on a variable input such as improved seed or fertiliser can have a big effect on revenue earned; $200 of personal expenditure produces no farm income at all even though it may add to the owner's satisfaction or improve his way of life.

Assume that a farmer is applying fertiliser (a variable cost input) to a hectare of crop. As the operating overhead costs per hectare are $6, the first $6 spent produces no income since it is used to meet overheads. The first dollar spent on fertiliser produces $3 of income, the next dollar produces $2.90, and so on. In Table 5.1 the operating profit obtained by applying extra units of fertiliser is shown. It is necessary to apply $4 worth of fertiliser before a profit is made. The farmer has to spend money on variable inputs in order to produce output and income and also in order to cover overhead costs.

The small farmer needs to produce a total gross margin per hectare which is $3 higher than the medium-sized farmer has to, if he is to just break even (Table 5.2). Assuming that both are operating their farm activities at the same level of efficiency, the medium-sized farmer will be able to make $3 more cash surplus per hectare than the small farmer. Often, when product prices drop and/or costs increase, the small farmer is unable to break even while his middle-sized counterpart can still show a surplus.

Table 5.2. *Relative total overhead costs per hectare of two sizes of farm*

	Small farm – 10 ha ($)	Medium-sized farm – 20 ha ($)
Living costs	600	750
Finance costs	100	180
Replacement of capital items	80	120
Taxes	45	70
Repairs to structures	20	40
Business expenses	15	20
Vehicle running costs	80	120
Total overheads	940	1300
Total overheads per hectare	9.40	6.50

The relative roles of variable and overhead costs in profit making are indicated when activity gross margins are calculated. A certain number of variable inputs have to be used just to produce enough gross margin to cover overheads. After this level, each extra unit of variable input produces profit, until a stage is reached where the cost of producing an extra dollar of income exceeds $1.

The difference between the total gross margin and the total overheads is the amount of money surplus the farmer has available for development, savings or extra personal or household expenditure.

Farm returns

Farm returns come from four main sources:

(i) Farm income, i.e. income from the annual farming or livestock operation. There are five broad sources or categories of farm income:

sales of crops, animals and animal products (milk, manure);

farm products consumed by the household(s);

dividends from cooperatives, farm related groups of which the farmer is a member;

non-cash income resulting from inventory change (extra stocks on hand at the end of the trading year);

off-farm work (such as share farming or contracting).

(ii) Household receipts, i.e. not derived from farming. For example: sale of handicrafts; profits from small trading.

(iii) Capital and machinery sales, i.e. sale of land, machinery or other capital items which are not a normal product of the year's business operations; such sales are not treated as part of the annual farm income.

(iv) Money from loans.

Details of various forms of farm return

The main source of income is normally the sale of crops, animals or animal products. By definition, the cash receipts shown in, say, cash flow budgets come from cash sales, loans, repayments by debtors. There can also be non-cash income resulting from (i) an inventory change, e.g. extra stock or animals on hand at the end of the trading year, or (ii) from farm products consumed in the home.

When we attempt to measure, in an economic sense, the amount of farm operating income earned in a given period, (usually 1 year), it is necessary to take account of changes in inventory. For example, take two farms of comparable size and efficiency which each produce 5000 kg of grain worth 10 cents/kg. The two farmers use the crops as follows:

	Farm A	Farm B
Sold	5000 kg	3000 kg
Kept	—	2000 kg
Total	5000 kg	5000 kg

Even though the cash income from Farm A is 5000 × 10 cents, and from Farm B 3000 × 10 cents, most economists would say that the true yearly income of both farms is 5000 × 10 cents. Grain kept in store is valued at the original selling price of similar grain. If, upon later sale, the actual price obtained differs from that shown as the value of the inventory, the difference is recorded in the next year's accounts as a profit or loss on storage. If there is a decrease in value of inventory

Table 5.3. *Returns and costs for semi-subsistence farm*

Returns ($)			Costs ($)		
1. Sales of farm products			*Total overhead costs*		
Crops	400		Living costs of family	500	
Animals	200		Wages, food, clothes for workers and family	250	
Animal products	100		Replacement of existing capital items	200	
Total		700	Taxes to village chief	100	
			Repairs to buildings	50	
			Costs of running motorcycle	100	
			Total		1200
2. Changes in inventory (stocks)			*Variable costs*		
Increase in cattle	+400		Seed, fertiliser, fuel, bags, sprays etc.		500
Decrease in stored grain	−250				
Net value		+150			
3. Value of saleable food eaten by family		800	*Finance costs* (can be included in overheads)		200
4. Household receipts			*New capital costs*		
Handicraft sale	60		New room on house		300
Profits from roadside stall	140				
Money from son working on building site	300				
Total		500			
5. Capital sales			*Household (personal) costs*		
Old plough	20		Included in overheads (above)		—
0.1 ha of land	100				
Total		120			
6. Money from loans			*Other taxes*		
Merchant	100		Tax on crop sales to marketing board		50
Bank	280				
Total		380			
A. Total household returns		2650	*B.* Total costs		2250

Difference $(A - B) = \$2650 - \$2250$
Surplus $= \$400$

over the year, the amount of decrease is deducted from the sales to give the true income. Thus:

	Value ($)
Value of inventory at start of year	500
Value of inventory at end of year	200
Change in value of inventory over year	−300
Cash sales of crop during year	1400
True income for year ($1400 − $300)	1100

Inventories can also be valued at the cost of production, but in farming this is difficult to assess. So, in measuring the true income of a farm, both cash sales and the value of the change in inventory are included as income items.

The value of produce consumed by the farmer and his family is also counted as part of farm income. It is income in 'kind', not in cash. It is usual to count only those items which have a market value. For instance, produce which was unsaleable because it had suffered some damage or because there was no market for it would not be counted.

When a farmer is a member of a cooperative society he will often receive annual dividends as rebates on purchases. These are accounted for under annual farm income. When a capital item such as land or machinery is sold, the money received is not part of the annual farm operating income. It is treated as part of the total receipts for the year. Money received from non-farm sources such as gifts from relatives, sales of handicrafts or work done elsewhere is not part of farm operating income. It is classed as household receipts. Whilst a loan brings money to the farm business, it is not counted as part of the income produced by the farm. Table 5.3 above contains the main elements of receipts and costs for a semi-subsistence farm.

When comparing cash returns and costs, both the value of (i) increased inventory and (ii) food eaten by family, have to be deducted from the returns. In the example above in Table 5.3, the increase in inventory and the food eaten by the family amount to $950.

When this sum is deducted from total household returns (A) of $2650, cash receipts are only $1700. When cash receipts of $1700 are compared to total costs (B) of $2250, it can be seen that there is a cash deficit of $550. This deficit would have to be financed from borrowings or from other sources (such as sale of unwanted capital items).

Questions

1 Give an example of: capital cost, total overhead cost, variable cost, finance cost, opportunity cost, hidden cost.

2 What are the 4 main sources of farm *returns*? Give an example of each.

3 Explain how we arrived at the following conclusion: 'For a short period, it can be worthwhile to continue production, even though the total costs exceed the total returns'.

6

Farm profits: financial statements and records

Profit

If you asked most farmers: 'How much profit did you make last year?' they would express their answer in terms of cash. The commonest answer would be 'I had only $300 at the start of the year, now I have got $700, so I suppose I made $400 profit'. He has kept his family for the year, maybe bought an extra couple of goats, pigs, or cows, got a $200 loan from the bank which he does not have to pay back until next year, won some bets, married a second wife, and sold an old bicycle. Accountants and economists would shudder at his definition of profit – nonetheless, the $400 gain is real to him; in his terms, it is profit.

There are almost as many definitions, as there are definers, of profit. For example, one person may define farm profit as the sum of:

(i)   the difference between the cash held by the farmer at the start and the end of the year;

(ii) the extra grain or animals on hand at the end of the year; and

(iii) the increase in value, over the year, of the assets the farmer owns, e.g. house, machinery and land.

Another person may use the rules of the local stock exchange in reporting the results of their year's operations. A third person may follow a definition conjured up by the particular school of economics of which he is a devotee, and so on. We will indicate where we stand, and stick to that definition right through. The reader can then make any changes which suit his own situation.

When speaking of profit, most people have some notion of the money left over from income, after all the costs which were involved in earning the income have

been deducted. Profit usually refers to some surplus or excess of income over costs (the net gain from a production process). To us, profit is the difference between the gross income and the operating costs. The operating costs are the sum of all the variable costs and the operating overhead (not total overhead) costs. Put another way, it is the total gross margin minus the operating overheads (see Chapter 5 for the definitions of these terms).

Annual surplus – a more practical measure

The amount of money which is left over at the end of the year after all 'costs' have been paid is known as annual surplus (or deficit). The annual surplus can also be calculated by subtracting from the operating profit any living or school expenses in excess of the operator's allowance, taxes, interest paid, loan repayments and any new capital investment income made. In Table 6.1 we illustrate the calculation of the annual surplus (or deficit) for a small commercial farm.

In this table both the true 'profit' of the business and how it is disposed of, are shown. However, it does not tally with the bank statement because gross margins can contain allowances for depreciation and for inventory changes. Thus farm animals may be replaced only every 4 years, yet an allowance is set aside each year so that funds are available for replacing the animals when they are old. A similar depreciation charge is made for plant and machinery. Inventory changes are added to, or subtracted from, stock values. Thus, if two calves worth $50 each are retained instead of sold, the farm has extra livestock capital of $100. This does not show on the bank statement as 'cash paid in'.

Table 6.1. *Calculation of annual surplus (deficit)*

	Amount ($)
Activity gross margins	
Crop A	200
Crop B	500
Dairy	160
Beef	100
Sheep	140
Other activities	20
Total gross margin	1120
Less 'operating overheads' (see Chapter 5)	700
Net operating profit	420

Total resource value of farm = $6000;
percentage return on resources =

$$\frac{420}{6000} \times \frac{100}{1} = 7\%$$

Disposal of operating profit	
Living expenses above operator's allowance	100
Income tax	70
Interest	60
Loan repayment	100
New capital investment from income	50
Total operating profit disposed of	380
Annual surplus	+40

Table 6.2. *Profit and loss statement*

	Activity A (crop) ($)	Activity B (livestock) ($)	Activity C (other) ($)	Total ($)
Receipts				
Sales of farm products	1000			1000
Livestock sales		1500		1500
Less purchases		−600		−600
Changes in inventory	100	−40		60
Contract work (off farm)			150	150
Total activity income	1100	860	150	2110
Less activity variable costs	500	400	20	920
Activity gross margin	600	460	130	1190
Less operating overhead				800
Operating profit				390

Presentation of financial results – farm financial statements ('accounts')

For some farms, especially larger plantations, cooperatives, and state farms, it is necessary to produce three financial statements each year so that the owners, members, or relevant government department will each have a clear picture of the financial health of the enterprise. The three statements are as follows:

(i) The profit statement (Table 6.2). This shows the profit or loss made from the year's farming or trading operations.

(ii) The statement of sources and use of funds (Table 6.3). This shows where the money came from and how it was used. This statement is helpful in drawing up the cash flow budget for the following year.

(iii) The balance sheet, or equity, or net worth statement (Table 6.4). This shows the value of the assets, estimated realistically according to market values at a given date, minus the debts. It is mainly concerned with the capital gains or losses of the farm business, and shows the net worth of the enterprise, or the owner's equity.

These items can be arranged, as in Table 6.4, to show briefly the current financial position of the farm enterprise. In the example shown, the farm is sound enough from the capital aspect, but current liabilities at $500 exceed current assets at $400. If his creditors pressed him the farmer might be obliged to sell some of his cattle to be able to settle his debts. When presenting such statements it is customary also to provide the corresponding figures for the previous year for comparison.

Return on capital (resources)

The term 'resources' refers to all the factors or means of production which are at the disposal of the farmer. The most important are: land; buildings; improvements such as built-up soil fertility and irrigation facilities; machinery; livestock; fuel; labour; management skills and credit. Depending on the conditions of land tenure, many of these resources can be converted to cash by selling them. The cash sum which would be

Table 6.3. *Statement of sources and uses of funds*

	Amount ($)
Sources	
Cash sales of farm products	1000
Cash sales of livestock	800
Deferred payments from previous years	200
Additional loans	1500
Additional sundry creditors	100
Sales of capital items	600
Other cash receipts	100
Total sources of funds (*A*)	4300
Uses	
Cash variable costs	600
Cash overhead costs	400
Capital investment	500
Interest	400
Loan and creditor repayment	200
Taxes	300
Living expenses	800
Increase in debtors	100
Other cash payments	50
Total uses of funds (B)	*3350*
Net funds flow for year (A − B)	950
Plus cash balance at start of year	600
Balance at end of year	1550

Table 6.4. *Statement of assets and liabilities*

	Amount ($)
Assets	
Fixed assets:	
Land and buildings	15 000
Total fixed assets	15 000
Working assets:	
Machinery	1500
Furniture	300
Cattle	600
Poultry	100
Total working assets	2500
Current assets:	
Grain on hand	200
Feed on hand	50
Commercial bank account	150
Total current assets	400
Total assets (*A*)	17 900
Liabilities	
Long term:	
Agricultural bank loan	5000
Total long term liabilities	5000
Current:	
Cooperative store account	200
Other creditors (e.g. merchant)	300
Total current liabilities	500
Total liabilities (*B*)	5500
Equity or net worth (*A − B*)	12 400

available from the sale (after paying off any debts owed on the farm) is the farmer's own capital. It is also known as his equity, or net worth.

If we express the farm's annual profit, after paying interest and taxes, as a percentage of this capital, the figure which results is one measure of the effectiveness of the management of the farm's resources. The percentage return on capital (equity) is calculated as follows:

$$\frac{\text{Annual profit after interest and tax}}{\text{Capital}} \times 100$$

Return on capital provides a guide to those responsible for the use of the capital (this could be an individual, a cooperative or a government department). It also allows the performance of this capital, invested as it is, to be compared with alternative possible investments. The farmer's capital (or equity, or net worth) is worked

out by adding the market value of all the resources the farm enterprise owns, and subtracting from that figure the total of all the money it owes. This is illustrated in Table 6.5.

A farmer's capital is sometimes expressed as a percentage of the total resources under his control. This is called the equity percentage and is calculated as follows:

$$\frac{\text{Resources owned} - \text{Money owed}}{\text{Resources owned}} \times \frac{100}{1}$$
$$\text{(assets)} \qquad \text{(liabilities)}$$
$$\text{(assets)}$$

$$= \frac{19\,500 - 6500}{19\,500} \times 100 = 66\%.$$

The main purpose of calculating the return on the farmer's capital is to show him how efficiently he is running the annual operations of his farm business. If

Table 6.5. *Calculation of farmer's capital*

	Amount ($)
Market value of resources owned (assets)	
Land and buildings (where owned by the farmer)	15 000
Animals	2000
Machinery	1000
Unsold products stored on farm	500
Money in bank account No. 1	1000
Total (*A*)	19 500
Value of money owed (liabilities)	
Money lender	1000
Development bank	3000
Cooperative	2000
Bank account No. 2	500
Total (*B*)	6500
Farmer's capital (or equity, or net worth, or A − B)	13 000

	Year 1 ($)	Year 2 ($)
Profit after paying tax and interest	650	800
Value of farmer's capital	13 000	18 000
Annual % return on capital	(650/13 000) × (100/1) = 5%	(800/18 000) × (100/1) = 4.4%
Increase in asset worth		5000
Net gain over year ($5000 + $150)		5150

the figure were only 1%, and other farmers with similar land, climate and capital were obtaining 4%, the farmer should ask himself the following questions:

Could I increase the rate of return on my capital by using better methods, borrowing extra money to improve production, or changing the mixture of activities on the farm?

Should I transfer my capital from this locality to a farm in a different locality where the return on capital is likely to be higher?

Should I transfer my capital out of farming altogether, and put it into some other form of investment, such as a transport business or a shop?

Is the return on my capital low because there has been a large rise in the value of my assets?

Should I use my increased net worth as financial backing to borrow more money, to further develop my farm, and so to earn more income?

The more rapid the annual increase in asset worth, the harder it is to maintain a constant rate of return on farmer's capital. For example, it is even possible for the percentage return to decrease, while the profit increases, due to a sharp rise in the value of land or of livestock. The following figures illustrate this:

An increase in asset value means that there is more collateral against which to borrow to use to help increase farm income. It can also mean that higher land taxes have to be paid, and (in some countries) more inheritance taxes, unless the title is passed on to children before the death of the owner. These consequences must be taken into account by the farmer and/or his adviser.

The course most commonly chosen by farmers whose return on capital is low is, first, to try and raise income by better management. If the returns are still low, then alternative uses for the capital should be investigated.

Although the rate of return on farmer's capital (equity) is one valuable measure of the use of capital on a farm, other measures also have some applications. A range of measures can usefully be employed.

Return on extra capital

This is one of the most important measures of capital use. In many cases, if additional capital is used to increase the existing farm resources, an increase in profits will result. When extra capital is put into drainage, irrigation or land clearing on an already existing farm, the farmer will want to know how well that extra capital has been used. One useful way of judging this is to express the extra net profit that is obtained as a percentage of the extra capital that was invested. Thus the rate of return on extra capital =

$$\frac{\text{Extra net profit after paying extra interest and extra tax}}{\text{Extra capital invested}} \times \frac{100}{1}.$$

In many cases, this figure can be as high as 50%.

If when, say, extra land clearing is being considered, estimates show that the rate of return on extra capital will be only 9–10%, the farmer should think very carefully about whether to go ahead with it or not. There is almost always uncertainty about the weather, prices and other happenings (such as floods, diseases and so on). Remember that he usually has to borrow at least some, if not all, of the extra capital needed to carry out the improvement project. The extra profit, after paying interest on the money borrowed and tax on the extra income earned, has to be high enough to cover the repayments on the loan.

As a general rule, it is unwise to invest in a new project on an existing farm unless the rate of return on extra capital is at least 10%. If the return on farmer's capital stands at 5% and the estimated return on extra capital of a project on the farm is 20%, then (as long as the project performs as expected) the return on farmer's capital will rise above 5%.

Calculating the return on extra capital is a simple way of testing whether or not a project should be adopted. This is very important when capital is scarce and has alternative uses. Other names for 'extra' capital are 'marginal' capital and 'added' capital.

Return on total capital

The market value of the total resources on a farm is often known as the total capital of the farm. It is calculated by summing the market value of the land and improvements (if relevant) and the value of the animals, machinery, tools, grain and fodder reserves. The rates of return on capital is calculated by expressing the annual operating profit as a percentage of the total capital. This is done as follows:

Percentage return on total capital (resources) =

$$\frac{\text{Operating profit}}{\text{Total capital}} \times \frac{100}{1} = X\%.$$

In calculating the annual rate of return earned by these resources, no account is taken of the debts owed. The calculated return then is the earning rate of the total bundle of resources employed in the business. In practice, the farmer has to manipulate the total resources under his control, not just the proportion

which is debt-free. If there is, for instance, a 30% debt on the property, it is not conceivable that a farmer would concentrate his efforts on only 70% of the area or on 70% of the animals on the farm.

Rate of return on total capital (resources) does not take account of interest, loan repayments, living costs, new capital and tax payments. All of these are specific to the financial and family structure, and to the standard of living of each farmer. Rate of return on total capital provides a guide to the earnings of the total resources as currently managed, unconfused by personal factors. It measures the following for each operator:

his effectiveness as a combiner of annual inputs, such as the labour, irrigation, fertiliser, seed and machinery resources;

the rate of earning of his capital – committed as it is to his farm – relative to the rate of earning of that capital if it were employed in some other income-producing field, such as transport or bank deposits (the real interest rate on government-backed bonds or the yield of safe investments shows the minimum relative fruitfulness with which capital can be employed in the economy).

Return on historical capital v current value

Historical capital usually means the value of the farm plus machinery and animals at the time of purchase or at incorporation as a cooperative or farming enterprise. It gives a misleading impression of the current earning capacity of the investment. For example, if a farm has an annual operating profit of $1000, an historical value of $10 000, and a current value of $20 000, then rates of return are 10% and 5% respectively. The rate of earning of the resources is measured more accurately by taking the return on the current value of total capital, because this gives a basis for comparison with the earning rate of today's value of the capital in other uses.

Capital gains

Increase in the capital value (resource value) may occur simply from a rise in the market value of land. It may also be increased by well-chosen investments in clearing land, supplying water, acquiring good breeding stock, for example. It is calculated by deducting the cost of capital investments made from the total increase in value of the assets over a given period. There is, however, no direct cause and effect relationship between capital investment and capital gain. For example, an expenditure of $500 on farm buildings does not necessarily mean that the value of the farm will rise by $500. They may not suit the plans of prospective buyers and may thus be valued by them at much less. It is often useful to measure capital gain in both 'real' and 'nominal' dollars (see Chapters 9 and 15 for reasons and details).

Records

Keeping track of what is going on requires some records. Keeping records which will be useful, can be useful. One part of good management is to have a recording system for the farm that will be used. Having accurate facts and figures is most useful when planning to borrow money, to get help from the government and for tax purposes. On commercial and state farms, records are necessary to produce the different financial statements (accounts) described earlier in the chapter. Farm records can also be used to help evaluate the enterprise mix on the farm. The farm adviser can often see where his farmer client is doing well, or poorly, if his past records are accurate. He can, to a small extent, use this information to help him to plan his future operations.

Records show what has occurred or is occurring; budgets tell what might be in the future. The only records worth keeping are those that will be used. Refer to Chapter 7 'Cash flows', to see how budgets and records can be used together for planning.

Records some farmers find useful

Farm diary
A small notebook in which to put down the key facts and figures of the business and farm activities, as they occur, is probably the most useful, practical (and often the only) form of record worth keeping. Farmers who do use a diary find that important facts and figures that could easily be lost or forgotten are permanently recorded for future reference (though they can be hard to find quickly).

Crop records
It can sometimes be useful to record what happens to each plot or crop each year, e.g.: type of crop; fertiliser application (time and amount); seeding; weeding; sprays; yields.

Livestock records
For management purposes, keeping track of livestock production and numbers can be worthwhile, e.g.: livestock production (fibre, milk and animals sold); livestock numbers (split into breed, sex and age groups); purchases; births; dead and missing; animals eaten by farm households; feed consumed by animals (on 'factory' farms).

Livestock inventory accounting
The main purpose of livestock accounts are to show net losses or gains in income, and to distinguish the increases or decreases due to changed numbers from the increases or decreases due to changes in market value. It is important to recognise the difference between the change in livestock value which is due to (i) change in numbers and (ii) market influences.

The change in stock value due to a change in numbers is a part of true farm income. If there are more stock this year than last year then the farmer has chosen to forgo some of the cash income from the extra animals by retaining some which are normally sold, i.e. he has invested his potential cash income in more animals.

Multicolumn records
Books with multicolumn sheets, can help in keeping farm financial records. Each column can be given a heading which fits the specific situation on each farm. Sheets with differing number of columns can be obtained. One farm situation may need only 8 or 9 columns, another as many as 27.

Below (Table 6.6) is a simple form of multicolumn sheet for payments with the first four columns used to identify and evidence the transactions. A more complex sheet may show variable costs for each enterprise, rather than lumping them together as in this example.

Table 6.6. *Simple multi-column sheet for payments*

Date	Details of transaction	Docket or invoice no.	Cheque no. or cash	Cheque or cash total ($)	Overhead costs ($)	Variable costs ($)	Capital costs ($)	Personal ($)
					Allocations			
2/3	Taxes	116	23	300	300			
10/7	25 bags fertiliser for crop	27	Cash	500		500		
20/7	Purchase of new plough	318	24	700			700	
30/7	Payment of school fees	65	25	400				400
Total payments				1900	300	500	700	400

Multicolumn sheets can also be used to:

record receipts from the different farm activities (as in payments, above);

record details of costs and income from each activity, e.g. for each crop, to show fertiliser, seed, spray, labour, water, machinery, harvesting, processing, transport and selling costs;

show how total cash wages bill was allocated among different activities;

show annual machinery depreciation.

A main feature of multicolumn sheets is that they can be adapted to many uses in farm financial records. They also allow each farmer to develop a system of records which suits his personal and financial situation.

Plant and improvement records
Examples of records that might be useful include:

costs and date of purchase or installation;

annual depreciation based on expected replacement cost;

insurance;

fuel use;

hours of use of machines;

major repairs and maintenance.

The record-keeping systems of most farmers consist simply of boxes full of bills and scraps of paper. It is rarely the oft-touted 'place for everything and everything in it's place' system. An orderly system can, however, reduce time spent on 'book work'.

Accounting for farm management purposes on commercial farms

Accounting for farm management purposes differ from standard accounting in a number of important ways, although certain basic accounting procedures are used in both systems. Farm management accounting aims to provide realistic information on the up-to-date state of both the farm business as a whole, and on individual activities. The data derived can help with forward planning, but only to a limited extent. Farm management accounting has a number of distinct features:

approximate market value, not historical costs, is the basis of valuation of most resources analysed;

inputs are related to outputs;

direct costing or the gross margin is the basis of activity analysis;

the production year, rather than the official year, is the basic accounting period (thus, depending on the industry, and the farm locality, the commencing month of the accounting year will differ);

physical records play an important part in preparing the statements, e.g. labour and machine usage, crop yields, feed consumption by various classes of livestock, sex and age composition of flocks and herds;

regular, up-to-date reports of actual progress against budgeted projections are given.

Farm records can be useful for measuring performance and for improving farm planning. Adequate records make it much easier to carry out a comprehensive analysis of the farm's performance, and are useful if applying for a loan. This is when physical and financial records come together. Both are needed for a critical appraisal of progress, and to make improvements in the farm finances.

Keeping full records just for the sake of keeping records is a waste of time and energy. A little time spent keeping a few orderly records, that will be used, is worthwhile. There are numerous cheap farm record booklets available from both extension and private sources in most tropical countries. We recommend that the farm adviser select the one(s) that best fit his client's needs.

Questions

1 What is your definition of 'profit'? Can you use this definition for farm management purposes?

2 Accountants and economists have different views of income, costs and profit. What are the key differences?

3 Operating profit is different to annual surplus. What is the difference?

4 What do we mean when we say 'costs are what you think they are'?

5 What does the profit and loss statement tell you?

6 What is a 'source and uses of funds statement'?

7 What does the net worth statement show? How would you use this information?

8 Why must both operating profit (return on capital) and capital gains be considered when investigating the profitability of a farm business.

9 What records do the local farmers keep? Do they provide sufficient information for you to do some farm management sums for them?

7

Cash flows

Introduction

The annual net cash flow (or net receipt) of money interests most farmers. For the semi-subsistence farmer, a budget also has to be drawn up showing the 'net food flow', i.e. the expected supply and demand for food, before the cash flow budget can be constructed. Both budgets can stimulate and assist farmers and their advisers to make plans.

The cash flow is found simply by subtracting the money spent over the whole year from the money received. Farm advisers often find it useful to calculate cash flows on a monthly or quarterly basis to help their farmer clients. These are especially useful for short-term crops. They may do this when they are budgeting next year's programme, or when comparing this year's actual results, with the expected results set out in the budget which they prepared before the start of the year. When a farmer is planning either next year's programme, or a longer-term project which may involve investing borrowed capital, he wants to know certain things from his adviser.

(i)   How much money or food the programme is likely to produce and how much the programme will cost?

(ii) When will he receive the money or the food, and importantly, when will he need money or food?

(iii) If the amount of money or food he expects to receive during a given period does not cover the amount he will need, how can he make up the difference? Will the bank provide a loan? Does the farmer have savings, or any stored reserves of food?

(iv) How much money will he be able to save in say, 1, 2 or 3 years' time if he undertakes this programme?

Cash flow budgets are important in:

planning farm development projects;

arranging loans;

preparing to cope with cash 'crises';

choosing between alternative farm plans;

comparing actual against budgeted results, so enabling corrective action to be taken on time.

Cash Flow Budgets

These are designed for the farm adviser to assist the farmer in planning his future activities, either on a short-term (monthly, quarterly or year-long) or a medium-term (4–5 year) basis. Cash flow budgets are also used to provide continuous feedback or monitoring during the period of the programme. The adviser will be able to compare the actual cash income and costs with the estimates made when the adviser drew up the budget. If the actual results differ greatly from the target results, then they can try to take steps to modify the situation before any serious harm results (see below).

The main feature of a cash flow budget is that it focusses specifically on cash. In contrast, other budgets, such as profit budgets, include non-cash items like inventory valuations and depreciation. For example, borrowing from the bank is regarded as cash received. Table 7.1 shows a cash flow budget for the first quarter of a 1 year period. The figures are hypothetical.

Comparison of actual and budgeted results

To help the farmer to check the actual financial progress of his plan against the expected or budgeted outcome, a budget layout, as in Table 7.2 is useful. It shows, for each quarter:

Table 7.1. *Cash flow budget for the first quarter of one year*

	1st quarter ($)	2nd quarter ($)	3rd quarter ($)	4th quarter ($)	Yearly total ($)
Receipts					
Sales					
Crops	500				2000
Livestock and products	200				500
Loans to be received	800				2000
Other receipts	30				200
Total receipts (A)	1530				4700
Payments					
Machinery cash costs	100				250
Variable production costs (excluding machinery)	150				650
Marketing costs	30				100
Overhead costs	70				400
Living costs	250				1000
Cash machinery replacement	0				0
Interest	100				250
New capital investment	150				500
Taxes	20				150
Other	30				200
Total payments before loans (B)	900				3500
Net cash flow $(A-B)$ (before loan repayment)	+630				+1200
Loan repayment	700				1000
Balance at end of period					
Credit					+200
Debit	−70				

actual quarterly receipts and payments;

budgeted or estimated receipts and payments for the quarter;

the total for the year so far, of actual receipts and payments;

the total yearly budget of receipts and payments;

the difference between the total yearly budgeted figure and the actual amount of money received and paid so far for the year.

Cash development budget

The next type of cash budget deals with a medium-term development which needs investment of borrowed capital, with a pay-off which will take several years. In Table 7.3 the main principles of cash development budgeting, which can be applied to any individual situation, are shown.

It is assumed here that the physical programme has already been planned. The cash flow budget simply tests whether the plan is financially feasible. Often, it is necessary to test three or four plans before making a final decision on the best programme. It will be seen that no repayments can be made on the loans during the first 2 years. In fact, the level of debt increases, as shown in the bank balance. This is a typical situation in a development programme. It shows the need for both banker and farmer to have good knowledge of the likely pattern of cash flow, if a development programme is to be properly financed. Many development programmes fail because insufficient credit is provided in the early years.

Over the 5 year period there is only a small

Table 7.2. *Quarterly comparison of actual against budgeted figures*

	Actual this quarter ($)	Budget this quarter ($)	Total accrued to date for year ($)	Budget for year ($)	Difference budget for year and total actual to date ($)
Receipts					
Sales					
Crops	400	500	400	2000	1600
Livestock and products	250	200	250	500	250
Loans to be received	870	800	870	2000	1130
Other receipts	20	30	20	200	180
Total receipts (*A*)	1540	1530	1540	4700	3160
Payments					
Machinery cash costs	120	100	120	250	130
Variable production costs (excluding machinery)	100	150	100	650	550
Marketing costs	30	30	30	100	70
Overhead costs	70	70	70	400	330
Living costs	300	250	300	1000	700
Machinery replacement					
Interest	110	100	110	250	140
New capital investment	100	150	100	500	400
Taxes	25	20	25	150	125
Other	25	30	25	200	175
Total payments (*B*) (before loan repayment)	880	900	880	3500	2620
Net cash flow (A − B) (before loan repayment)	+660	+630	+660	+1200	+540
Loan repayment	660	900	600	1000	340
Bank balance at end of period					
Credit	0			+200	+200
Debit	0	−270			

reduction in debt. The rate of reduction of debt over the next 5 year period would be faster. For a detailed discussion of development budgeting see Chapter 14.

Food and cash flow budget

On semi-subsistence farms, the proportion of crops, animals or animal products consumed by the household usually exceeds the proportion sold. For such a farm the equivalent of the cash flow budget of the commercial firm is the 'food and cash flow budget'. It has exactly the same purposes as the cash flow budget:

to indicate the likely balance between supply and demand of the items in which the farmer is most interested;

to help the farmer to plan measures to cope with a period of deficiency (examples of seasonal food and cash flow budgets are illustrated in Tables 7.4 and 7.5).

Table 7.3. *Total cash flow budget of a farm on which a development programme is proposed*

	Year 1 ($)	Year 2 ($)	Year 3 ($)	Year 4 ($)	Year 5 ($)
Cash receipts					
Sales of farm products	2000	3000	4500	5000	5500
Loans to be received	8700	1500			
Other receipts	500	300	600	500	500
Total receipts (*A*)	11 200	4800	5100	5500	6000
Cash payments					
Variable costs	500	750	1000	1250	1500
Overhead costs	1000	1100	1200	1300	1400
New capital investment	8000	1000	500		
Interest	400	500	500	450	400
Living costs	1000	1100	1200	1300	1300
Taxes	300	350	400	500	550
Total payments (*B*) (before loan repayment)	11 200	4800	4800	4800	5150
C – Net cash flow (*A* – *B*) (before loan repayment)	0	0	+ 300	+ 700	+ 850
Loan repayment	0	0	300	700	850
Bank balance at end of year (cumulative):					
Credit					
Debit	8700	10 200	9900	9200	8350

Table 7.4. *Food and cash flow budget for a semi-subsistence farm*
(a)

	1st season (kg)			2nd season (kg)			3rd season (kg)		
	Supply	Demand	Difference	Supply	Demand	Difference	Supply	Demand	Difference
Grain crops	100	160	− 60	100	170	− 70	300	180	+ 120
Root crops	400	200	+ 200	100	190	− 90	200	200	
Vegetables	200	100	+ 100	250	150	+ 100	400	200	+ 200
Tree crops	100	150	− 50	300	150	+ 150	20	150	− 130
Animal products	20	30	− 10	20	30	− 10	45	35	+ 10
Overall adequacy of food supply for household needs	Adequate: can substitute some roots for grains			Inadequate: little room for substitution			Adequate: small amount of substitution		
Proposed plan of action on basis of budget	Sell some surplus roots			Buy grain and meat Try to raise more animals by this time Hunt Borrow money Seek more off-farm work			Store some surplus grain Sell surplus grain and vegetables		

(b)

Season	Expected cash receipts for surplus products (A) ($)	Expected cash payments for deficient products (B) ($)	Difference ($) (A − B) +	−
1	25		25	
2		60	−	60
3	130		130	

Table 7.5. *Cash receipts and payments budget for a semi-subsistence farm*

	1st Season ($)	2nd Season ($)	3rd Season ($)
Cash receipts			
Crops, animals and animal products	25		130
Off-farm work	15	18	10
Other cash receipts (not loans)	5		3
Essential loans needed		78	
Total cash receipts (*A*)	45	96	143
Cash payments			
Farm expenses	12	15	25
Family	20	80	30
Interest and Taxes	4	6	3
Other cash payments		4	
Total cash payments before repaying loans (*B*)	36	105	58
Net cash flow before loan repayments (A–B)	9	−9	85
Money available to repay loans (if necessary)	9		85
Money			
Owed		78	
Saved	9		7

Questions

1 Construct a food and cash flow budget for next product year for a semi-subsistence farmer that you know. What does it tell you about the borrowings which will be needed?

2 Draw up a cash flow budget for a commercial farm. Is this budget likely to prove of any help when you come to discuss the financial programme with the banker?

3 What contingency plans have you made if the budget does not turn out as expected (either worse or better)?

4 Can you see any practical value in keeping a 'budget versus actual' cash flow sheet?

8

Gross margins

Introduction

The gross margin of a farm activity is the difference between the gross income earned and the variable costs incurred. It is probably the most commonly used measure in farm analysis and planning, and also is now used widely in extension publications. For a farm on which several different activities are carried out, the total gross margin is the sum of the gross margin from each activity. In any one year, total gross margins should not be less than total overheads if the farmer is to avoid extra borrowing, or the sale of some of his assets and/or use of his cash savings.

Calculating activity gross margins

Gross income minus variable costs is a straightforward calculation. There are, however, some slightly different methods used to work it out. There are two components of gross income:

sales of produce;

change – increase or decrease – of inventory (the term 'inventory' means what stocks there are, or were, on hand).

Thus, the grain inventory refers to how much grain (maize, sorghum etc.) the farmer had, or has, stored away. If he had 500 kg stored at the start of the year, and 300 kg at the end of the year, the change in inventory of his grain was:

$$-200 \text{ kg} \times \text{(say) } 10\text{¢/kg} = -\$20.00$$

Similarly, if he had 5 cattle at the start of the year, and 8 at the end, the change in his livestock inventory is:

$$+3 \times \text{(say) } \$400 = \$1200$$

There can be problems when measuring gross income. The income from sales of produce is not always received immediately or all at once, as some products are sold to marketing boards or cooperatives which work on a 'pool' system. Often payment from the 'pool' for the delivery of 1 year's crop may extend over 2–3 years. Furthermore, on semi-subsistence farms, some of the product is eaten by the family, some is sold.

The simplest way to calculate gross income, where gross margins are being used for planning, and for comparing alternative future activities, is to assume that all of the product from the activity is sold, and all the income is received at once. This gets over the problem of whether it is sold, eaten, stored or not fully paid up for 2–3 years.*

When using gross margins for analysing the existing activities on a farm, then it is necessary to define activity gross income more completely and precisely. Here, there is a need to take account of:

cash sales of (say) the crop;

amount of saleable crop eaten;

how much was stored rather than eaten or sold (i.e. change in inventory);

money owing from the board or cooperative after the first payment was received when the crop was delivered.

Thus, the true activity gross income for 1 ha of a grain crop may be as follows:

---

*If the delay in receiving full payment for a particular product were significantly longer than this (and especially if inflation is high), and if this activity was being compared to an activity where the income from sales was received in full, immediately, then the above method of using gross margins for planning and comparison could be a bit misleading as to which is the 'best'.

Production	2.5 tonnes/ha
Price	$120/tonne
Gross income if all sold	$300
Disposal of crop	
Sold for cash	$120
Eaten by household	$ 80
Extra grain stored	$ 40
Future payment from pool	$ 60
Total	$300

Table 8.1. *Simplified calculation of gross margin per hectare of cash crop*

	Amount ($)
Income:	
Assumed sales	280
Variable costs:	
Seed	10
Fertiliser	25
Additional paid labour	45
Repairs and maintenance of machinery	15
Fuel and oil	15
Sprays	11
Insurance	7
Transport to market	21
Selling costs	17
Total	166
Total gross margin/ha	114

Gross margins for crops

An example of a gross margin calculation for a crop is given in Table 8.1. Variable costs of crop activities are made up mainly of:

pre-harvest or growing costs seed, fertiliser, water, extra labour, sprays, direct machinery costs (fuel, oil, repairs);

harvest costs extra labour, direct machinery costs, bags;

marketing costs direct costs of storage, processing, transport and selling.

Gross margin from animals

The gross income (not gross margin) of an animal activity is made up of sales of animals, animal products, by-products, and inventory changes. The four main components of the variable costs of any animal activity are:

(i)   Feed, which includes the costs of forage crops, hay, straw, silage, purchased feed, home-grown grains, maintaining improved pastures, payments for grazing, as well as direct labour costs.

(ii)   Husbandry, i.e. medicines, cleaning materials for milking sheds, and veterinary services.

(iii) Marketing, i.e. transport, processing and selling.

(iv)   Breeding and replacement stock where not reared on the farm.

In many livestock activities, the change in value of the flock or herd due to change in numbers (inventory changes) is an important part of the yearly 'profit'. Increase in value due to increased stock numbers must be included as income because, if these extra stock were sold, there would have been a greater cash return. The farmer has chosen to forego this extra cash income to increase his livestock numbers. So, he has really invested this amount in next year's livestock capital. Conversely, any reduction in livestock numbers (inventory) from the start to the end of the year, must be deducted from cash receipts to give the true income.

The value of animals in an activity could also change due to market influences. Including the change in stock value which is due to market influences, as well as changes in numbers in gross income can sometimes distort the analysis and reduce its usefulness as a planning tool. It is better to use the same values per head at the end of the year as were used at the beginning. Periodic stock revaluations can then be used to account for market influences on stock values.

Gross margins are worked out per head and per hectare, however, for non-breeding activities. For breeding activities gross margins are best worked out per breeding unit, or per herd unit. That is, the gross margin includes income from animal offspring, as well as a sire or semen cost. The gross margin is the difference between the total income derived from the breeding activity and the total costs associated with that breeding activity. The gross margin of any animal activity can be calculated as follows:

(i) Sales of animals and animal products, plus value of animals and their products eaten.

(ii) Sales of by-products (e.g. manure).

(iii) Increase or decrease in value of stock due to change in stock numbers from the start to the end of the year, valued at price of similar stock at start of year.

(iv) (i + ii + iii) = gross income.

MINUS

(v) cost of feed, husbandry, and marketing.

(vi) Cost of (replacement) animals bought.

(vii) (v + vi) = total variable costs.

(viii) (iv − vii) = gross margin.

Below is an example of this method of working out gross margins. *We will use a 10 cow dairy herd, which also had 2 unmated yearling female replacements at the start of the year. For simplicity assume 100% calving, 1 cow death/year.*

GROSS INCOME

(i) *Sales of animals and animal products*

(a) Animal Sales $ $

1 cast for age (old) cow	@400 =	400
5 bull calves	@ 70 =	350
2 heifer calves	@ 80 =	160
1 CFA bull every 4 years ($\frac{1}{4}$ × 500) per year	@500 =	125
Total animal sales	=	1035

(b) *Animal product sales ($)*
Milk from 10 cows = 5000

(ii) *Sales of by-products ($)*
Cow manure = 40

(iii) *Change* in Total Value of Stock over the Year due to increase in numbers:

3 extra heifer yearlings @200 = 600

(iv) Gross Income (i + ii + iii) ($) = 6675

VARIABLE COSTS

(v) Cost of feed = 2500
husbandry = 120
marketing = 175
Total = 2795

(vi) *Cost of replacement animals bought ($)*
1 cow (to replace one which died) @500 = 500
1 bull every 4 years, @600 = $\left(\frac{600}{4}\right)$ = 150

Total = 650

(vii) *Total variable costs ($)* (v + vi) = 3445

(viii) *Total gross margin ($)* (iv − vii) 6675 − 3445 = 3230

Gross margin/cow unit = $323

In Table 8.2 a simpler method of calculating gross margins is given.

Table 8.2. *Gross margin for beef*

Gross margin for 10 cow unit beef herd		
A. *Income – sales/year*		
6 yearlings @ 300	=	1800
2 cows (old) @ 400	=	800
1 bull (every 4 years, @ 600)	=	150
		2750 (A)
B. *Variable costs*		
Husbandry @ $5.00/cow	=	50
Feed costs @ $30/cow	=	300
Bull replacement (1/4 years @ $800)	=	200
		550 (B)
C. *Gross margin (A − B)*	=	2200 (C)
Gross margin/cow	=	220

It should be noted that it does not include inventory changes.

The way to allocate the wages of permanent workers and depreciation of machinery, plant and buildings directly involved in an activity, is always a problem. If it would lead to better decisions about which crop activity should be chosen, then these costs should be allocated. In most cases, the wages of permanent workers and depreciation of machines are costs which must be met and remain roughly the same, regardless of the activity mixes chosen. Then they need not be considered in deriving activity gross margins for use in planning.

Use of gross margins in planning

The gross margins per hectare of crop and per head of livestock are widely used for comparative analysis of activities on one farm, and between farms in similar environments. For example, if the gross margin on one farmer's maize crop is $150 per hectare and the average gross margin of neighbouring farmers is $300, then he should seek advice on the technical reasons for his own poorer economic performance.

Valid comparisons can only be made in terms of a production unit common to all of the farms or activities being compared. This unit can be the land area, if the land used by each activity is equally suitable. It could also be per unit of labour, per $100 of capital invested, or per breeder unit or per head of livestock. In tropical countries where the family labour available is an important factor (or the village labour available, in the case of a cooperative enterprise) then gross margin per unit of labour may be the most appropriate base for comparison.

Note that the gross margin technique assumes a linear relationship as the activity is expanded, i.e. if the area of maize is doubled, we assume that the gross margin for the extra hectares will be the same as for the original area. This is not always so, as there can be a diminishing returns effect as the activity is expanded. While in many cases it is reasonable to assume a linear relationship when planning to increase the area, the farmer and his adviser should keep the possibility of diminishing returns in mind as the activity is expanded (see Chapter 4).

For simple changes in activities, if one activity has a gross margin of $100/ha on a particular soil type, and another has a gross margin of $200/ha, then, provided certain qualifications (which will be discussed) are met, the activity with the gross margin of $200 should replace the one with $100.

Gross margins (GMs) are also a useful first step in deciding on the best combination of activities on a farm. The procedure here is to select the highest GM per unit of the most common limiting resource (hectare of land, $100 of capital, rotation, man hour or man week of labour), and expand it until some other restraint is met. Then the activity with the highest GM of all the remaining available activities is introduced until it too meets a restraint, and so on.

For example, a limiting resource on the farm might be arable land (though on small farms, the first restraint may be seasonal labour). Vegetables may be selected first because they have the highest GM. They are expanded until the number of hectares planted is such that the vegetable activity meets the restraint of, say, a lack of reliable seasonal labour to harvest the crop quickly enough to avoid losses of yield and/or quality. A crop activity with a different harvest period should then be introduced. (Alternatively, there may be a limiting supply of labour for weeding.)

The activity to select is the one with the highest expected GM of the alternative crops which could be planted at this time. For an example, we will assume that maize is a crop which meets this criterion, though the maize GM per hectare may be less than for vegetables. The maize will then be planted on as much of the remaining arable land as is possible until the next restraint is met.

We next assume that at the time of the year when he is planting maize the farmer draws near to the limit of the next restraint – working funds. So he plants as many hectares of maize that his working funds will allow, and waits until he gets more working funds perhaps from the sale of his vegetables. With these funds he is able to plant the remainder of his arable land with that crop (from the range of crops which can be grown at this time of the year) which has the highest expected GM. The procedure can be repeated with the animal activities on the non-arable land until all the available resources are used to the limits of the restraints imposed on their use.

We noted above that the use of GMs was a useful

first step in planning the best combination of activities on a farm. In practice, it is probably the most commonly used technique for planning activity mixes. A more sophisticated approach is to use substitution ratios, which are discussed for growing a mixture of crops in Chapter 11. Using GMs is a simple and quick way to plan changes in activities, activity mixes, or to analyse a farm business. There is usually no need to consider overheads, and the sensitivity of the proposed change to possible variations in yield, prices or costs can be readily tested.

Application to semi-subsistence farms

In planning the food supply on semi-subsistence farms, where there is a relatively low cash input, the main objective is to produce an adequate diet for the household. The contribution which the main activities make to the total food supply is often more important than the size of the GMs. Of course, if one or two crops have very high GMs, it may be best to grow them for sale and to buy some food. If most of the activities are concerned with food supply rather than with cash sale, then activities should be chosen on the basis of the starch and protein they will supply at different times of the year.

Although 'food supply' has thus replaced the 'gross margin' of the commercial farm as an objective, exactly the same principles in choosing activities and activity mixes can be applied. The adviser to the semi-subsistence farmer should identify the activity with the safest, highest expected 'food supply' and advise expanding that first until it meets a restraint. Then the next best activity is expanded to the point where it in turn meets a restraint. On semi-subsistence farms, there can be many restraints, e.g. labour (both in numbers and skills at various critical times of the year), money for essential cash inputs, and also rotation restraints where the land must be fallowed after a certain period of cropping. If one of these is the major restraint then that should be used as the starting point for GM analysis instead of land.

The use and usefulness of gross margins for planning

Calculation of GMs is a simple and direct technique, and a useful first step in any form of farm budgeting and planning. A farm adviser needs only to identify production and planning constraints, and to budget incomes and variable costs for each activity. Some care needs to be taken when using GMs as a basis for planning. For example, it is frequently found that cash crops show the highest GM per hectare. Before the crop area is increased it is necessary to know:

the maximum area that can be grown with present soil type, hectarage, labour, working capital and machines;

the technical limits of expansion (e.g. under a fallow or legume crop rotation system, it may only be possible to cash crop a particular field for 3 years out of 6).

There are obvious physical and financial limits to expansion such as availability of suitable land, shortage of both labour in peak periods and of credit. In parts of West Africa, for example, the availability of women to do the planting and harvesting, sets a limit to rice growing.

It is very important to investigate the technical efficiency of present practices before GMs are used as a basis for making changes. A decision to replace a crop with a seemingly low GM should be preceded by a check to see if its gross margin could be raised by better use of fertiliser and pesticides, by greater attention to performing operations on time, or by better field management. Other circumstances under which it would not be wise to adopt the activity with the highest GM would be where suitable labour to handle the activity was not available or where capital was limited. To illustrate this latter limitation, let us take a farmer who has access to only $600 credit and who has to decide between two alternatives, *A* and *B*, with the conditions indicated below:

	Activity *A* ($ per ha)	Activity *B* ($ per ha)
Gross margin	30	20
Capital needed	60	30

If he adopts activity *A*, the farmer needs to invest $60 for every hectare. Since he has access to only $600 of credit for capital, he can only have:

$\dfrac{600}{60}$ or 10 hectares of activity *A*.

This will give him an annual GM of 10 ha × $30 or $300. If he adopts activity *B*, he can have

$\dfrac{600}{30}$ or 20 hectares of activity *B*.

This will produce an annual total GM of 20 ha × $200, or $400. Thus it is useful to consider both the gross margin and the capital needed per hectare, before deciding which activity to choose.

GM analysis has limitations. Intermediate activities such as using crop stubble cannot be included separately in the analysis. Also, it arrives at a plan which gives highest returns to the assumed most limiting resource; in fact, that resource may not be the most limiting.

There are some more sophisticated farm planning techniques such as linear programming. Linear programming uses returns to all resources in the analysis and for choosing between activities, and is based on the principle of substitution.

Questions

1 What factors do you need to put numbers on (i.e. to have detailed information) in order to calculate the gross margins of a cropping activity? An animal activity?

2 Can you easily get all the information you need from the local farmers to do some gross margins analysis?

3 Do you have a good idea of the most common levels of yields, prices, work rates and input costs for your area?

4 Is it correct that if activity A has a gross margin of $40 per hectare, and activity B a gross margin of $20 per hectare, you should always expand activity *A* and reduce activity *B*? Why is this a 'trick' question?

5 List the ways in which you would use gross margin calculations in helping a commercial farmer to plan. What about a largely subsistence farm family – how would you use GMs to plan their farm activities?

6 What are the main limitations to how useful a tool GM analysis can be in farm analysis and planning?

9

Time is money

Introduction

Managing time and its effects is another important task for the farmer. First, there is a cost, in terms of income forgone, if he fails to complete certain operations on time, e.g. sowing, harvesting and disease control. Second, a dollar received today is usually valued more highly than a dollar received in, say, 6 years' time. The reason is that today's dollar usually can be used productively, and it will grow to more than a dollar in 6 years' time, just by putting that dollar in the bank and leaving it there. Here, we will deal with the second meaning.

There is the further issue of inflation. Inflation also makes the dollar in the pocket worth considerably more than the expected dollar of next and later years; we will handle this point in some detail later.* But even if there were no inflation, it would be better to have a dollar today than one in 6 years' time. The reason: you can invest today's dollar, and if you reinvest it (and the, say, 5% interest it earns each year) it will grow to $1.34 in 6 years' time.

The correct answer to the problem of whether to use land, labour, and capital in one way which might bring in a large net cash flow, but not for a number of years, or in another way which may offer the prospect of a smaller net cash flow but sooner, involves putting a value on money received and spent at different times in the future. The decision on the best policy is made easier by using the technique of discounting the cash

*People *also* get differing amounts of money and/or satisfaction from spending their time in different ways. So, another aspect of the 'time is money' issue refers to the value a person may attach to using their time in one way rather than in another. For example, some use an airplane when there is a train or bus available to perform the same transport service. We will not elaborate on this aspect of 'time' here.

flows to their net present values (NPVs). This allows valid comparisons to be made between alternative investments which have differing flow patterns of costs and returns in future years. So, the two relevant points, when thinking about the role of discounting in farm management decision-making, are:

the number of years in the future; and

comparing alternatives.

Let us take the situation where a farmer, (or any investor), after looking at a few development options, has already decided, for various reasons, that one particular programme is best for him. It will be a 'number of years in the future' before the programme reaches its full potential (reaches a 'steady state'). But, since he has already 'compared alternatives', and has firmly committed himself to a course of action, the technique of discounting for project appraisal has no relevance for him.

It is only when he is choosing between several available farm development options that discounting has a place. Here, he and his adviser need to have a valid way of comparing alternative programmes when the amounts of money spent and earned are different and occur in the future.

Cash flow development budgets are the basis for appraising whether or not a project should be undertaken. Details of the use of cash flow development budgets are given in Chapter 14. Here we concentrate on the 'time' aspects of development budgeting. This chapter needs to be used together with Chapters 14 and 15 to fully learn how to do development budgets.

To compare alternative medium- to long-term farm development plans, it is necessary to understand both compounding and discounting; they are based on a similar concept. As a farm management tool, discounting is usually of greatest value when the investment period exceeds 3–4 years. Note: We wish to stress again that compounding and discounting are useful concepts, even if there were no inflation in the economy.

Compounding and discounting

Compounding

The technique of compounding is best explained by an example. Suppose that $200 is invested at a compound interest rate of 10%. What will be the value of the $200 in 5 years' time? At the end of the first year, there is still the $200. There is also $20 interest which the $200 has earned. Over the next year 10% interest is paid on $220. Every year the interest earned can be reinvested and earns 10%. After 5 years there is a total of $320, i.e. $200 today will be worth $320 in 5 years' time, provided that the interest is retained and reinvested at 10%. The formula for calculating the end value of a compounding investment is:

$A = PV(1 + R)^n$, where

A is the future amount to which the investment will grow;

PV is the present value (or today's value) of the sum invested;

R is the interest rate expressed as a decimal; and

n is the number of years for which the investment is made.

Discounting

Discounting is the opposite process to compounding. When discounting, today's (or the present) value of a sum of money spent (or received) in the future is calculated. The formula for calculating the present value of a future cash flow is:

$PV = A/(1 + R)^n$,

where the symbols have the same meaning as in the compounding formula. Note that the discounting formula is simply a rearrangement of the compounding formula.

Thus, the present value of $320 received in 5 years'

time, assuming an interest rate of 10%, is $320/(1.10)^5$ where $(1.10)^5$ means 1.10 multiplied by itself five times; i.e.

$1.10 \times 1.10 = 1.21$
$1.21 \times 1.10 = 1.33$
$1.33 \times 1.10 = 1.46$
$1.46 \times 1.10 = 1.61$

$PV = 320/1.61 = \$200$ (approx.)

The discount tables given in Table A (in appendix 1) have been calculated on this basis. They show the present value of $1 discounted at different interest rates for different periods of time. Similarly, the compound interest tables show the future value of money invested today at compound interest for various periods.

Discounting tables

To use discounting tables, once a discount rate has been chosen, the present worth of an amount due in any year in the future is found by multiplying that amount by the discount factor shown in the tables for the corresponding year. Using the following excerpt from the discounting tables we can calculate the discounted value of a $500 per year net cash flow for an interest rate of 10% for each of the 3 years.

At 10% the discount factor for the first year is 0.9091. This means that each dollar of the $500 which the investment earns on the last day of the first year will be worth only 90.91 ¢ in present day, 'real' terms, i.e. $500 × 0.9091 = $454.55. To find what the $500 earned in the third year is worth in present day terms, look up the 10% discount factor for year 3, which is 0.7513. The answer is $375.65. In total, the 3 year cash flow of $500 each year is worth $1243.40 in net present value (NPV) terms.

For this example, where the same sum was involved for each year, we could have worked out the present value using *annuity* tables. An annuity is a constant amount to be received or spent at regular intervals at a given discount rate. Reading from the present value of an annuity table (Table C, Appendix 1) the annuity factor (3 years, 10%) is 2.4868. $500 each year × 2.4868 = $1243.40 (see Table 9.1).

Present values

Having looked at how the NPVs are calculated, let us look at how they are used in investment analysis and

Table 9.1. *Present value of a future lump sum at 10% discount rate*

Period (years)	10% Discount factor		
n			
1	0.9091		
2	0.8264		
3	0.7513		
4	0.6830		
5	0.6209		

Period	Cash Flow	× Discount factor	= *PV* (today's value)
Year 1	500	0.9091	454.55
Year 2	500	0.8264	413.20
Year 3	500	0.7513	375.65
Total (NPV)			1243.40

what they mean. Important: in the following discussion we are talking about 'what to do' in a situation where there is no inflation. (Later the implications, and complications, of inflation are discussed.)

In the following example (Table 9.2) two farm projects, A and B, each have a life of 5 years and require an initial investment of $1500. Net cash flows, after deducting a minimum personal living requirement, but not including interest payments, are as follows:

Table 9.2. *Net cash flows for two competing projects*

Total undiscounted value ($)	Year				
	1	2	3	4	5
Project A	+900	−900	+600	+600	+600	—
Project B	+1500	−1500	—	—	—	+3000

Note: in Project A, $600 'profit' was earned in Year 1. So the net cash flow in Year 1 was +$600 − $1500 = $900.

These cash flows are in terms of today's (not future) dollars. In each of Years 2, 3, 4, project A is expected to bring in (net) 600 of today's dollars. Project B costs

$1500 in Year 1, and returns $3000 in Year 5. A sum of money in the hand today could be invested at a rate of earning (interest) and it would grow to a larger sum in each year in the future. Or, put another way, a sum of money received in the future, has the same value as a lesser sum today; the latter could be invested to grow to the future sum.

It is necessary to know how much these sums of dollars received in the future from each project are worth in the form of sums of money in the hand today (NPV). The project which produces the net cash flow which gives the biggest amount of dollars in the hand today is the one which is best (if you want to make the most money from the funds you are going to invest). A negative NPV means that the project is *not* earning at the interest rate used for discounting. Zero NPV means that the original capital invested is fully recovered and has earned a rate of return equal to the discount rate. A positive NPV means that the project is earning at a rate greater than that of the discount rate used.

To calculate how much these future dollars are worth in today's terms (remembering today's dollars can be invested to grow to bigger future sums) it is necessary to discount those future sums back to present values. To do this, a decision has to be made about the interest rate at which to 'discount'. The rate to use depends on:

the opportunity interest of the capital the farmer has, or could borrow, i.e. the rate it could earn in other acceptable investments (both on or off the farm, if the farmer was willing to consider both investment options);

the farmer's personal attitude to different investment opportunities, e.g. whether he is prepared to accept 10% interest as a reasonable return on his money, or whether he wants more than 10%;

his preference for an extra dollar soon versus an extra dollar he might receive much later.

Ideally, farm investments should return as much interest (or 'profit') as would be obtained if the funds were spent on alternative investments of comparable safety and growth prospects.

In this example, with no inflation, the market rate (which is also the real rate) of return for a fairly 'safe' money investment may be 3% (possibly the best

Table 9.3. *Present values for project A ($)*

	Net present value ($)	Present Values ($) (Years)				
		1	2	3	4	5
0%	+900	−900	+600	+600	+600	—
5% Discount Factor		0.95	0.90	0.86	0.82	0.78
NPV at 5%	+693	−855	+540	+516	+492	—
10% Discount Factor		0.91	0.83	0.75	0.68	0.62
NPV at 10%	+537	−819	+498	+450	+408	—

Table 9.4. *Present values for project B ($)*

	Net present value ($)	Present Values ($) (Years)				
		1	2	3	4	5
0% Discount	1500	−1500	—	—	—	+3000
NPV at 5%	+915	−1425	—	—	—	+2340
NPV at 10%	+495	−1365	—	—	—	+1860

alternative use of the $1500). The projects A and B are more risky than this so we will discount them by real interest rates of 5 and 10%. The present values of the projects are shown in Tables 9.3 and 9.4.

The data in the tables illustrate two fundamental points about project appraisal:

the project with the highest net present value is the most profitable;

the discount rate chosen can affect the ranking of projects.

For example, at 0% interest, Project B is better than A. This is still true at 5% discount rate. However, when high real discount rates are used, say 10%, then Project A is better than Project B.

Inflation

Inflation is an expansion of the supply of money in relation to the supply of goods and services; in consequence this leads to a decline in the amount of things the money can buy, i.e. to a decline in the value of money. Rates of inflation vary country by country. In making predictions about the future, and building the effect of inflation into budgets, it is necessary to distinguish 'nominal' from 'real' money values.

Real and nominal values

A 'real' value is the value of an item (e.g. 1 kg of yams) measured at a given point in time. The value of the item, on this definition, does not change over time. On the other hand, a 'nominal' value of an item such as 1 kg of yams may change from its original value as time passes. Nominal money value is the dollar value over time, in terms of the actual face value at each particular point in time. Nominal money values are not directly comparable with any other dollar value at some other point in time if the purchasing power of the dollar has changed. Real money value is the dollar over time, with the effect of changes in purchasing power removed. Real money values are directly comparable.

Nominal money values are expressed in terms of the actual present-day price of things we have to purchase. A loaf of bread may have cost 25¢ 6 years ago: today, the nominal or cash price of a similar loaf of bread is 75¢. However, in real terms, (in terms of the purchasing power of 25¢ 6 years ago), the price of bread may not have risen at all. It is simply that 75¢ of today's nominal money can only buy the same amount of goods and services as the 'real' 25¢ did 6 years ago.

The most common measure of the degree of inflation is the 'consumer price index'. Periodically, the price of a standard package of basic goods and services required by consumers is recorded. Changes in this composite price from its price in a given base are expressed as changes in an index number. If the number changes from 100 to 105 over a year, the rate of inflation has been 5% for that year. This does not mean, of course, that the prices of all the products the farmer sells, and of those he must buy, will have inflated at that rate. Each is subject to differing conditions and differing rates of inflation. For instance, the price of products exported to world markets, and of imported supplies such as fertiliser, will be subject to very different influences from, say, the cost of local labour.

Inflation is a fact of economic life. Losses in buying power are likely to remain an important feature when budgeting developments where payoffs occur in the future. In the example below, the effect of a 10% inflation on real income is shown. Even though there has been an increase of 300 nominal dollars over the 5 year period, in real terms there is a loss of \$80.

Effect of 10% inflation on real income

	1984	1989
Net income (nominal or current dollars)	700	1000
Net income in 1984 (constant or real dollars)	700	$1000 \times 0.62^a = 620$
Loss in real income	0	$700 - 620 = 80$

[a] The 0.62 is the 10% discounting factor which converts 1989 dollars to 1984 dollars. It is found in Table A (discount table), Appendix 1.

We will show how to handle inflation in farm development budgets. Also, with an example, how to take account of the fact that it is better to have a dollar now, than one in 5 years' time, even if there were no inflation. An important consideration is the effect inflation has on (i) the interest rate to use in discounting, and (ii) the future sums of cash.

Interest rates

The interesting aspect of the concept of the 'time value of money' is the way it is incorporated into budgets for investment decisions. The 'time value of money' is expressed as a percentage of the sum of money, and is called the interest rate. Interest is the percentage rate at which a sum of money grows. This rate of increase, say 10%, can be made up of a real component and a nominal component, e.g. if inflation was 5%, and a sum of invested money earned 11%, then about 5% of this would be due to inflation of the money, and about 6% would be real gain.

In detailed project analysis, where projects are being compared, interest has to be taken into account. This is done by calculating the net present value of the different projects. The value today of the future net cash flows is calculated by multiplying the net cash flows in each year by a discount factor, based on one or more chosen interest rates.

Whilst the calculations are simple using discount tables, the tricky bit (once the physical and financial details have been used to work out the expected cash flows of the different projects being compared) lies in

Table 9.5. *Net cash flows (before interest charged) for two projects with similar capital needs (\$)*[a]

	Year 1	Year 2	Year 3	Year 4	Year 5
Project A	−200	−100	+200	+300	+400
Project B	−500	0	+100	+350	+650

[a] Both have NPVs of 600.

deciding which interest rate to use (i.e. what the real rate of interest is which could be earned). For comparing two projects, it is simplest to use real (not nominal) dollars of the cash flows and then discount by real interest rates: the implied assumption here is that the relations of real costs and prices will not change. Inflated values could be used and discounted by market interest rates which include an inflation allowance. The end result (the NPVs) will be the same.

Where the availability of funds is not a major limiting factor, say on a well developed farm, the relevant discount rate is the current market interest rate. A farmer with limited funds on a part developed farm is in a different situation. His on-farm investment may earn a much higher return on capital (say 15% after tax). The discount rate he would apply to off-farm investments to evaluate them would be 15%. In other words, he has to apply an 'opportunity' interest rate of 15% on his capital if he is contemplating putting funds into an off-farm investment project. Let us now work through some examples of project appraisal using discounting.

Examples The expected annual cash flows, in real (constant) dollars for two projects, each of which require the same amount of capital investment, are shown in Table 9.5.

The net present values for the two projects, using real dollars and real discount rates of 10% and 20%, are as follows:

Discount rate	NPV	Project A (\$) Year 1	Year 2	Year 3	Year 4	Year 5
0%	+600	−200	−100	+200	+300	+400
10%	+337	−182	−83	+150	+204	+248
20%	+185	−166	−69	+116	+114	+160

Discount rate	Project B ($)					
	NPV	Year 1	Year 2	Year 3	Year 4	Year 5
0%	+600	−500	0	+100	+350	+650
10%	+261	−455	—	+75	+238	+403
20%	+71	−514	—	+58	+168	+260

Table 9.6. *Net cash flow for project A (real dollars)*

	Year 1	Year 2	Year 3	Year 4	Year 5
Cash in	300	450	800	950	1100
Cash out	500	550	600	650	700
Net cash flow	−200	−100	+200	+300	+400
Total	−(200+100)+(200+300+400) = 600				

So, project A 'wins', whether 10% or 20% is used – but it is a draw if 0% is used. Let us have a look at project A in more detail. The net cash flow (NCF) figures used for project A were based on constant (or real) dollars. They were derived as follows (Table 9.6):

Having decided on the best project, it is now necessary to do a finance budget to see how many nominal, not real, dollars will be needed to carry it out. If both prices for products sold (cash in) and costs of inputs bought (cash out) inflate at the same rate, say 15% (this assumption about 'same' is not very true), then the future cash flows in nominal (face value) dollars, would be as in Table 9.7. The nominal dollar figures are found by referring to the 'growth at compound interest tables in Appendix 1.

Inflation is expressed in compound terms, i.e. 15% inflation means that the nominal value of the currency next year will be 15% higher than this year's value. Thus, the compound interest tables can also be used to calculate growth in nominal dollar values due to inflation.

The inflated cash flow figures of Table 9.7 can be used for a finance budget which the farmer and his adviser will need to show to the banker when applying for a loan for farm development.

Thus, the farmer needs $362 from the bank for the first 2 years, and he will (hopefully) have $304 to repay this loan in Year 3. In Years 4 and 5 the net cash flow is so good he will even be able to lend to the bank. Note that in nominal dollars, the net cash flow is $1271 (cf., $600 in real dollars). The 1271 nominal dollars in net cash flow have only the same purchasing power as 600 real dollars, when there is a 15% inflation rate over each of the 5 years of the project.

As mentioned earlier, it is not very realistic to assume that prices of products and costs of inputs will increase in value (inflate) at the same rate. It is more likely that farm costs will increase at a rate faster than prices for farm products will. Using the costs and prices

of Table 9.6 above, we will assume that prices increase at 10% (compound) and costs rise by 15% per year. This is shown in Table 9.8.

With different inflation rates for costs and prices, the project has a net cash flow in nominal dollars of only $605, compared with $1271 when inflation rates were similar for costs and prices.

The 605 nominal dollars net cash flow from the project have a lower purchasing power than the 600 real dollars with no inflation, because costs have increased at a higher rate than prices. If there were no difference in the inflation rate of costs and prices, the farmer would be no worse off in real terms whether there was inflation or not, except perhaps for having to pay higher taxes. In the example shown in Table 9.8, inflation makes the farmer worse off than if there were no inflation. In real life, when there is inflation, it is usual for farm costs to rise faster than prices for farm products.

Discounting in theory and practice

In theory, discounting should be used for comparing all investment decisions where expenditures and revenues occur over a period of time. In practice, many farm investments are evaluated on the basis of partial budgets, return to marginal capital, and cash flows. These all have the advantage of avoiding the complexity introduced by discounting, and the disadvantage of ignoring the effects of time on cash flows.

As the time effects on cash flows are but one of the many unknowns involved in an investment decision, it is difficult to say just when to use the discounting technique and when to follow the rule that 'near enough is good enough'. There seems one obvious situation where to ignore time would not be 'good enough', and that is where the cash flows occur over a long time period (we say more than 3–4 years for the projects being compared).

Table 9.7. *Cash flows adjusted for inflation ($)*

	Total	Year 1	Year 2	Year 3	Year 4	Year 5
Net cash flow in real terms	600	− 200	− 100	+ 200	+ 300	+ 400
Compounding factor to correct for 15% inflation		1.15	1.32	1.52	1.75	2.01
Net cash flow in nominal dollars	1271	− 230	− 132	+ 304	+ 525	+ 804

Regardless of whether time effects on money are included in an appraisal, the cash flows are crucial. It is best if the investment generates cash flow surpluses early in the life of the investment – with uncertainty, the sooner the investment pays for itself the better. As well, the higher the end value, or salvage value, of the investment, including any expected gains in asset values, the more attractive will be the investment proposal. Finally, the best decision will depend on the farmer–investor's objectives, and how he feels (his intuition) about the options of spending his funds in particular ways.

When to use real and nominal dollars in budgeting

When comparing two or more projects with a view to finding out which is better (or best), then in choosing the numbers to use in the budgets, simply use the present, or today's, values for costs and prices. This is because, if inflated nominal values are added to real values, the answer is meaningless. It is like adding loaves of bread and numbers of fishes and calling the resultant sum 'bread–fish' which does not mean anything to anyone. By using present (or today's) value in budgets to compare projects we are, in effect, using real prices and costs, so that the resultant budgets will at least be expressed in common terms.

When discounting, the NPVs can be calculated either by using real cash flows and real interest rates of discount, or by using inflated future cash flows and discounting by an interest rate which includes an allowance for the inflation. The same answers will result whichever way the sums are done. The important point is to be sure not to carry out some 'hybrid' sort of analysis, such as using real dollars and discounting by market rates of interest which include an allowance for expected future inflation.

However, once a project has been chosen, after having been compared with others, then the cash flow budget should be expressed in nominal (i.e. inflated) dollars, because if it were presented to a financier in real terms, it would understate the amount of nominal dollars needed to fund the project in future years.*

As we showed above, the present values of the net cash flows over time are calculated by discounting at the relevant interest rate; the project with the highest NPV should normally be the one selected. Even though the actual financial outcome of the project, in inflated currency, will be different from the budgeted result using today's prices, the basis for selecting the project is sound.

Three ways of judging projects

Three different methods are commonly used to help farm investors to decide between the relative profitability of alternative investments. The first uses net present value (NPV), which is simply today's value of future revenues, less the present (or today's) value of the costs involved in producing the revenue. The second method uses the internal rate of return (IRR) which is that interest rate which just balances the present values of cash receipts and cash outlays or, (the same thing) makes the total net cash flow equal to zero. The third method is called benefit–cost analysis (BCA).

The NPV alone tells us nothing about the pattern or the timings of the cash flows. And, depending on the farmer's psychological make-up, and the present state of his finances, he may have a stronger preference for dollars received in the first few years, which are closer to the present, and may seem more 'certain', than for

---

*Using the same reasoning, the effects of inflation have to be taken into account when planning for the future replacement of machinery (see Chapter 13).

Table 9.8. *Cash flow, in nominal dollars, with different inflation rates for prices and costs*

	Total NCF	Year 1	Year 2	Year 3	Year 4	Year 5
Cash in						
Prices in real dollars	3600	300	450	800	950	1100
Inflation factor at 10%		1.1	1.2	1.3	1.5	1.6
(A) Prices in nominal dollars	5095	330	540	1040	1425	1760
Cash out						
Costs in real dollars	3000	500	550	600	650	700
Inflation factor at 15%		1.15	1.32	1.52	1.75	2.0
(B) Costs in nominal dollars	4490	575	726	912	1137	1400
(C) Net cash flow (A − B) in nominal dollars		605				

dollars which he will receive in the more distant future.

With the internal rate of return (IRR), an investment is considered worthwhile if its IRR exceeds the interest cost of using the capital to invest in this way and not in another. The IRR is similar to the annual average return on capital over the life of the project, adjusted for the effects of time. It can be seen as the maximum interest which a project could pay for the capital used, if it is to recover all the capital, and breakeven. Assuming that investments have similar capital requirements, the investment with the highest IRR is the most 'profitable'. Using the IRR as a method of investment appraisal means that there is no need to specify an interest rate before doing the calculations. However, without a modern calculator, it is a more awkward procedure than the net present value method, because one has to use a number of 'trial and error' interest rates before the net present value of cash receipts and costs can be balanced, i.e. made equal to zero.

With discounting methods of assessing profitability there is an implicit assumption that the present or future values are lump sums which can be manipulated over the life of the project, by borrowing and lending at different times, at market interest rates, in order to generate the stream of cash flows which is desired. In practice, capital markets are not perfect enough to ensure that this will be possible. For this reason, as well as the size of the NPVs and IRRs, it is useful to look at the pattern of the cash flows.

Usually, both the NPV and the IRR method will rank alternative investments in the same order of profitabil-ity although, in some circumstances, NPV and IRR will rank two projects differently. Then it is necessary to compare the difference between the two NPVs, and relate these differences to any different amounts of capital used to generate these differences in the NPVs of the respective projects.

Whilst there are difficulties with NPVs and IRRs as decision aids, as long as the decision-maker appreciates the theoretical basis and limitations of these techniques, they can be a useful and usable guide to decision-making.

The benefit : cost ratio is another way of using the discounting technique in project appraisal. In the form of a simple ratio, it relates the present value of total benefits to the present value of total costs. It is found by dividing the sum of the present values of 'cash flows in' (benefits) by the sum of the present values of the 'cash flows out' (costs). The example in Table 9.9 illustrates the procedure.

For farm purposes, the benefit : cost (B : C) ratio can be a guide to the worth of a scheme. Using a discount rate of the value placed on one's own money (e.g. if a 10% return is needed, then discount at 10%), where the benefit : cost ratio is greater than 1, the scheme is profitable; if it is less than 1 the scheme is unprofitable. A difficulty with using B : C ratios is that ratios alone tell nothing about the size of the sums involved. Even though the differences in the B : C ratios can be relatively small, the difference in net present value can be large.

From the figures in Table 9.9 the benefit : cost ratio

Table 9.9. *Benefit : cost (B : C) ratio of farm development project at discount rates (30% and 10%), assuming no borrowing ($)*

		Year							
		1	2	3	4	5	6	7	8
Benefits	Total								
Cash received		0	455	1365	2050	2733	3188	3644	3644
Salvage (terminal) value[b]									9200
(B) Benefits (Undiscounted)	26 279	0	455	1365	2050	2733	3188	3644	12 844
(C) Costs (Undiscounted)	13 990	1100	1920	1690	2070	2250	2060	1550	1350
Present value of benefits @ 30%	5107	0	268	621	717	737	637	583	1541
Present value of costs @ 30%	4894	847	1132	760	707	607	412	248	162
Benefit : cost @ 30%	1.04								
Present value of benefits @ 10%	14 038	0	373	1024	1394	1694	1785	1858	5908
Present value of costs @ 10%	9210	1001	1574	1267	1407	1395	1153	790	621
Benefit : cost @ 10%	1.52								

[a] Summary: B : C @ 30% = 1.04; @ 10% = 1.52.

[b] Salvage values: most development programmes involve investment in items such as plant, livestock buildings, irrigation equipment and land clearing. If the project reaches a development 'steady state' situation in, say, 8 years, then one way of appraising it is to assume that it is sold. If so, then the plant, building and stock which were bought have a terminal or 'salvage' value. The cash value of these assets is treated as income received in the year that they are 'sold'. See, for example, above, where the salvage in Year 8 of the improvements was $9200.

changes from 1.04 to only 1.52. But the NPV (which is the difference between the PV of Benefits and PV of Costs) increases from $213 to $4828. The small change in the B : C ratio occurs despite a huge change in discount rate from 30% to 10%.

One common use of B : C ratios is to appraise government projects such as flood mitigation and dams for irrigation schemes. In this instance, the costs assumed are those of the government (or the taxpayer) in providing the development. The benefits are the increased profits able to be made both by individuals and government utilities (such as added freight for railways) as a result of the development. A 'favourable' B : C ratio is one where $1 invested by the government gives a return of over $1 to the beneficiaries of the scheme after an interest charge (viz, the discount rate) has been applied.

Forecasting

The major problem when evaluating an investment proposal is to make useful estimates of the future level of costs, prices and yields. Budgets are usually based on the expected, 'most likely', level of costs, prices, inflation and yields – and of course it is most unlikely that things will actually turn out exactly as expected. In some way this uncertainty about what will happen in the future must be included in the evaluation of the project. This can involve testing the effect of some possible levels of prices, yields and costs on the net result of the investment, i.e. a parametric budget, using conservative estimates of the benefits, or a higher discount rate, or by including some contingency plans for the 'if this happens' situations.

The only way to handle costs in a budget, unless there is special knowledge about future changes in input costs, is to assume that they will continue to rise at the same rate in the future as they have during the past 2–3 years. Price forecasting, by its very nature, is a hazardous and uncertain operation. Again, there is a need to use real money prices when comparing projects and nominal money values once having chosen a project.

In some countries, there are considerable amounts

of funds and skilled talents devoted to estimating likely demand, quantities and prices for major commodities. Most of this work is done by government sources and marketing boards, and also by some universities and private traders. These are all probably a better source of information on likely prices and costs than are the 'two Gs' ('gut feeling' and 'guesstimation').

Yields

The farmer's own experience, and that of local farmers on comparable land, who use similar technology, together with a 'fink factor' – i.e. an extra allowance to take account of likely developments in new yield- or quality-increasing technology (frequently deriving from local experiments) – seem to be the only bases on which a reasonable estimate of likely crop yields and animal performance can be made.

Inflation

Predicting the rate of inflation accurately is also very hard. The market rate of interest is made up of percentage 'real' return which investors hope to achieve, the expected rate of inflation, and an allowance for risk, e.g. with 10% inflation and 2% real return expected on the next best investment, then, without considering risk an investor will want to have a chance of getting at least 13% on his money from an investment proposal.

Productive values

The discounting formula can be used to put a value, called the productive value, on any resource which can be used to produce a flow of income over time. Assume that an asset is expected to produce an operating profit of $200 each year. If the asset has a market value of $2000, then the asset is earning at a rate of

$$\frac{\$200}{\$2000} \times \frac{100}{1} = 10\%$$

return on capital. Another way to think about this is the following. If a lender wants to get a return of 10% on his money (say the next-best alternative investment offers 9%), and an investment (an asset which generates income) is expected to produce an operating profit

of $200 per year for many years into the future, then the lender can afford to pay

$$\$200/0.10 = \$2000$$

for the asset, and still get a return of 10% on his money. Thus, if the operating profit of a farm being examined is $30 ha and 8% is the ruling rate for investments of comparable safety, we can say that the current productive value of the farm equals:

$$\$30/0.08 = \$375/\text{ha}.$$

In some areas, agricultural land values are so high that even good managers have difficulty in maintaining a rate of return of more than 5% from farm operations. In these areas, personal, social or capital growth considerations may lead the owner, or potential buyer to accept a lower rate of return than he could earn on the money market. When calculated this way, productive values merely reflect the expected future profits from using the asset for a long time into the future, and do not take account of that other major element which goes to make up market prices, which is the expected capital gain.

Caution should be exercised when capitalising operating profit. The following should be taken into account:

small changes in the operating profit have large effects on capitalised or productive value;

the interest rate used has a great effect on the total capital value calculated e.g. operating profit $30, interest rate 8%, productive value $375/ha; operating profit $30, interest rate 12%, productive value $250/ha; operating profit $40, interest rate 12%, productive value $333/ha;

the capitalised 'productive' value refers to the value of the farm in operating condition, i.e. land, stock, plant and improvements (thus, to get the value of the land and fixed improvements, the value of the stock and plant have to be deducted from the capitalised value/ha);

most authorities on land valuation have ruled against productive valuation in favour of market value, though the former is useful in estimating a theoretical value of the land based on its income earning capacity from farming.

There is not necessarily any relation between the market price and productive value of land. In fact, they are rarely the same, largely because expected future capital gains also go to make up market prices.

Perpetuities

Where positive net cash flows from a project are expected to continue for many years (30–40) into the future, the present values of these later flows can be calculated by using a technique known as 'discounting for a perpetuity'. It is described in specialist works on discounting, such as J. Price Gittinger's *Compounding and Discounting Tables for Project Evaluation*, published by Johns Hopkins in 1973.

Questions

1 The difference between nominal and real money values, and real and nominal interest rates, is crucial:

  Distinguish between nominal and real terms.

  When should you use real dollars and when should you use nominal dollars in project appraisal?

  When do you use real interest rates?

  When do you use nominal interest rates?

2 What is compounding?

3 What is discounting?

4 What is the difference between discounting to take account of the opportunity cost of funds and discounting to account for inflation?

5 With current inflation levels in your country, what do you estimate the real rates of return on government bonds to be?

6 List the pros and cons of NPV, IRR and BCA as measures of project profitability.

7 How would you choose what figures to use for yields, costs and prices when you do not know what they will really be, and you do know that they vary quite a lot?

10

Planning changes

Budgeting techniques

Now we turn to the crux of the decision process – the choice between alternatives. The purpose of this chapter is to show how choices can be made systematically so that the selections made are more likely to prove successful, even when risk is involved.

We can identify two broad types of change – simple and 'complex'. Simple changes involve, for instance, deciding whether to replace tomatoes with beans, or changing over from producing eggs to poultry for eating. In the case of the 'complex' change, extra capital investment is involved. Examples of such a change are putting in a well and pump for irrigation, clearing extra bush land or buying machinery. Usually, it takes several years before the new plan is working properly and reaches a regular pattern of costs and receipts, called the 'steady state'.

Both situations are described here. Where a simple change is contemplated, it is not necessary to go through all the steps suggested. Before a farmer/or his adviser can make a choice between two or more alternative farm plans, they need to know how the plans compare in terms of 'profitability'. If some small adjustment to the present farm organisation is being proposed, the structure and arithmetic of the budget will be very simple. On the other hand, if a complete reorganisation of the farm business is under consideration, the budget may involve many calculations.

Some basics

The first need in preparing reliable budgets is a sound knowledge of the technical aspects of agriculture and of the circumstances of the individual farm. It is no use dreaming up numbers to put in the budget and then expecting the answer to mean anything. Relevant, simple farm records can sometimes be a help in quantifying the present position with regard to the use of inputs or labour, machinery, fuel, seed, fertiliser, feed, and so on, in relation to the levels of output of crops and livestock of the present systems. This information will need to be supplemented with technical data from other sources such as neighbouring farms and experimental stations, and with the observations of an experienced extension worker. Then estimates must be made of the input and product prices which are expected to apply to the future planning period.

Changes in farm activities usually have a major impact on costs and there will be changes in the composition of the gross income. Budgets need to be drawn up for the present and the proposed farm plans so that the profitability of the two systems can be compared.

First, select one or more possible farm plans for which budgets are to be prepared. The resources available to run the operation should then be listed (for example, areas of land of different type or condition, the amount of capital available, buildings and labour). Next, the possible types of production should be considered in the light of the restrictions on the plan imposed by personal factors and fixed resources. Thus the fertility and topography of the land may exclude some types of production, and the choice of some alternatives will be further restricted by limited supplies of labour or capital.

From this general review of the possibilities, the broad outlines of one or more potential farm plans will emerge. Each such plan should first be specified in physical terms (i.e. the rotation to be adopted and the number and types of crops and livestock to be farmed and details of their requirements). The technical feasibility of the plan can then be confirmed, before it is put into financial terms.

When using budgets to evaluate courses of action it is very important to keep 'the figures' in proper perspective. Whilst budgets are an extremely useful tool in making decisions, there is often a strong temptation and tendency to read too much into 'the numbers', to regard them as being in some way predictive or 'true'. The numbers are (or should be) the best information available, at the time the decision has to be made. It is important not to fall for the trap of reading too much into this information. Nevertheless, budgeting is probably the most powerful tool which can be used to improve decision-making on a farm, so we shall discuss the logic and procedures involved in some detail.

In the preceding chapters we showed the place of some basic planning tools: the three most important were gross margins, return on marginal capital and cash flow budgets. Here, we will introduce a most useful technique, partial budgeting, and also describe whole farm, break-even, and parametric budgeting. We concentrate on budgeting in the most common, and important, planning and decision-making situation facing any farmer – whether or not to adopt a new activity, or to modify an existing one.

Deciding on a new or changed activity

Before a farmer adopts a new activity, or changes an existing one, some or all of the following points need to be taken into account by him and/or his adviser:

market prospects for the new or extra product;

physical and technical aspects of making the change;

change in 'profit', assuming 'normal' seasons and prices (this usually requires a partial budget);

risk and variability associated with the change – see pp. 85–94 (this is such an important consideration that we have devoted the whole second part of this chapter to it);

amount of extra capital investment needed to bring about the change;

return on extra capital invested;

net cash flow, over time, which is expected to result from the change;

expected change in assets and debts over time;

human and social aspects;

intangibles.

On a semi-subsistence farm, the contribution to the family's food supply and the degree of risk are of prime importance. These points should be looked at systematically to permit valid comparisons between alternative changes. For a simple change in activities without extra capital investment, only market prospects, technical aspects, change in profit and risk need be considered. Details of each point are explained below.

Market prospects

The farmer has to know when and how he would sell the new product (resulting from the change). His adviser needs to find out, from as many sources as possible, the likely market and the range of possible prices for his product, especially at the time it will be ready for sale. With some products in a small market, a large increase in production could cause prices to fall. As far as is possible, the farm adviser should also find out what the likely increase in supply of the product from other producers would be. After giving due weight to these factors, he should be in a position to make a reasonable guess about the prices that can be expected. This point is dealt with further in Chapter 11.

Physical and technical aspects

The physical aspects to consider are the resources available (labour, land, soil type, animals, tools and machines) to carry out the change, and whether extra resources will have to be obtained. Technical requirements may include specific types of fertiliser, animal husbandry techniques, and new methods of harvesting and marketing. Avoiding large peak seasonal workloads may be important for farmers with a fairly fixed supply of labour. The labour force (family and hired) needs to have the skills and knowledge required by the project. If it does not, it may have to acquire them. Many projects which look attractive on paper before they are started, turn out to be failures because the farmer and the workers lack the ability to carry them out.

To ensure that all relevant aspects of the change are taken into account, it is best to draw up a detailed plan of the land, machinery, tools, fertiliser, water, labour

Table 10.1. *Conventional partial budget format*

A – Losses with change	B – Gains from change
Increase in expenses ---	Decrease in expenses ---
Decrease in revenue ---	Increase in revenue ---
Total (*A*)_____	Total (*B*)_____
Net increase or decrease in 'profit' (*B* − *A*) = ————	

Table 10.2. *Whole farm profit budget*

	$
Gross margins (GM)	
20 ha of sorghum @ $175/ha	3500
50 sheep @ $20 per sheep	1000
50 goats @ $22 per goat	1100
10 beef steers @ $120 per head	1200
12 beef cows @ $140 per head	1680
Total gross margin	8480
Overhead expenses	
Taxes	150
Registration, licenses	75
Insurances	130
Light and power	60
Office expenses, accountant's fees	52
General freight, cartage	85
General repairs	150
Depreciation of plant, improvements and buildings	400
Wages of one permanent person	1000
Pick-up running costs	700
Operators allowance	1100
Total overhead expenses	3902
Operating profit = (GM − overheads)	4578
Capital	
90 ha @ $625/ha	56 250
20 ha cropping plant @ $170 capital/ha	3400
Livestock – sheep 2000	
– cattle 2000	5000
– goats 1000	
Total capital	64 650

$$Return\ on\ total\ capital = \frac{Operating\ profit}{Total\ capital} = \frac{4578}{64\,650} \times \frac{100}{1}$$

$$= 7.1\%.$$

[a] Hired labour, seeds, fertilisers, sprays and feed have been considered when working out the gross margins.

and skills needed, expected yields, and number and types of animals. Allowance has to be made for inefficiency in the early stages of a change, especially if the farmer is unfamiliar with the new activity. If the physical and technical bases of the operation are not sound, its economics will also be unsound.

Change in profit
The change in profit is a key test of the merit of the proposed change. The usual calculation for this is a 'partial budget'. The word 'partial' means that attention is focussed only on those costs and revenues which would alter as a result of any change made.

A partial budget is used in planning a proposed change, within the overall plan, and only shows the extra expenses and the extra revenue resulting from the change. The net profit or loss can be expressed as a percentage of the extra (or marginal) capital involved, thus giving a preliminary basis for comparing the percentage returns on capital which could be earned in alternative projects (as long as the comparisons are in the same-value money, i.e. real or nominal terms – see Chapter 9). The conventional method of presenting a partial budget is given above (Table 10.1). (We find that it can be rather confusing to use, even for people who have had some training; we present an easier method later in this chapter.) Two main categories, (*A*) and (*B*), must be considered.

Whole farm profit budgets for commercial farms
In the whole farm profit budget, probable future annual operating costs and gross income of a farm business are set out. Every item of cost and gross income is determined according to the farm plan. It is useful to make the calculation in terms of gross margins and overhead costs. This shows the contribution each activity makes to operating profit and also

makes it easier to study the effect of changes in activity levels on profit. The gross margin and overhead expenses for a hypothetical farm are shown in Table 10.2. The farm has five activities and an activity budget would be prepared for each activity, in order to determine the gross margins to be included in the complete budget.

Two cases in which it is useful to prepare a whole

farm budget are when a major reorganisation of the farm business is contemplated, and when a programme for a new farm is being planned. For example, the conversion of a cattle and mixed cropping venture to a dairy farm may be under consideration. It is obvious that annual feed and fertiliser requirements will be much different. Investment in plant and labour requirements will probably also be different.

Capital Aspects
Even if the plan is physically possible and is expected to lead to an improvement in profit, a loan may be needed to put it into effect. The amount of extra capital needed should allow for any sales of plant, livestock, or other assets no longer required if the plan is executed. The standard layout for this calculation is shown on Worksheet 4 (p. 000) and is known as a capital budget. The interest charge for the use of the borrowed capital is an important item of extra cost.

Percentage return on extra capital
The percentage return on extra capital is another important measure of the merit of a proposed change. It is calculated by expressing the expected increase in 'profit' as a percentage of the extra capital put into the change. As it can often take 3 or 4 years before the new activity is stabilised and working properly, the extra profit figure used in the calculation should be that obtained when the extra income becomes fairly stable. For example, if the profit before the change is $1000, and the profit after the change is $1500, as shown in Fig. 10.1, the decision-maker would choose $1500 as the basis for calculating the extra profit figure of $500.

It is simplest to do the calculation in today's dollars, not in terms of the inflated dollars which might apply in the future steady state, i.e. in real terms, not nominal terms. The return then is a real return, comparable with other real returns from other investments (see Chapter 9). If the extra capital invested was $2500, then the rate of return on extra capital would be:

$$\frac{500}{2500} \times \frac{100}{1} = 20\% \text{ (real rate of interest)}$$

Note: it is important to look very closely at any figure which is called return on marginal capital.

It is a tricky question as to whether the return on marginal capital should be calculated before interest

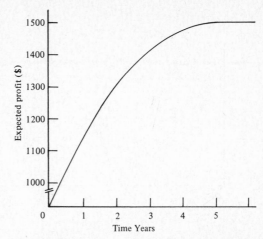

Fig. 10.1. Expected 'profit' after a change in farm plan is made. The proposed change generates a stable level of 'profit' by about Year 5.

costs or after interest costs are deducted. Both can be informative, so long as it is made clear what the rates of return measure. Either way, the return on marginal capital can be compared with the equivalent real return which one could hope to get from an alternative investment.

Imagine a sum of money, made up of personal money and borrowings. If used in a certain way this sum promises to earn at a rate of 20% per year. This figure of 20% can be compared with the expected earnings from other uses of the *same* funds. However when a number of different projects are being looked at, each involving different sums of money (different proportions of which consist of personal and borrowed money) then the rate of return after interest is the more informative figure and is the figure with which to compare the alternative uses of the capital. The return after interest represents the return to the investors own capital.

In practice, in most development projects, the farmer has to borrow money and interest payments are an important 'cost'. For this reason, in the example which follows we have deducted the interest costs and the final figure is return on marginal capital after interest has been paid.

To sum up, you cannot compare returns on marginal capital from two alternatives if (i) they are not both in the same values (real or nominal), or (ii) if one return is before interest and the other is after interest has been

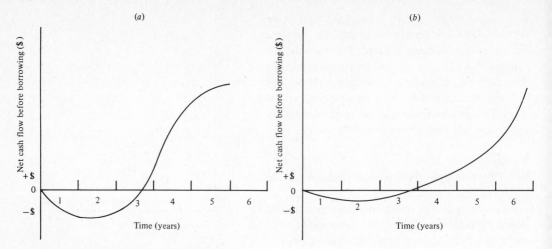

Fig. 10.2. Expected annual total farm net cash flows before
borrowing, from the adoption of two different
developments:
(*a*) development A; (*b*) development B.

deducted. To compare return on marginal capital with
the rate of return from an alternative investment using
the same funds, the return on marginal capital before
the deduction of interest is usually used. For projects
involving different amounts and proportions of own
and borrowed funds, rates of return after the deduction of interest are usually compared.

In most situations it is not attractive for a farmer to
invest in a new farming development unless the rate of
return on marginal capital, after interest, is at least
10% in real terms to cover:

the risks involved;

the need to pay back loans from the extra profit;

the returns available from other uses of the capital.

The methods of calculating the expected rate of
return on extra capital described above are a useful
screening device for selecting developments. If the
percentage return is high, then the development war-
rants further study and a cash flow budget. If it is low, it
can usually be rejected, and an alternative studied.

Pre-loan net cash flows

Even though two developments may show a similar
rate of return on an extra capital investment of say
$2500, this does not mean that they are equally
attractive to a farmer. For example, the total farm
annual net cash flows before borrowing (cash received
minus cash spent in a year, excluding loans needed)

may differ between two possible developments (see
Fig. 10.2).

The cash flows shown in Fig. 10.2 are those before
borrowing, i.e. they indicate the amount of money
which has to be borrowed in the early years if the farm
is to remain solvent. Development A has large cash
deficits in the early years, but high positive cash flows
from about the third year on. Development B has only
a small deficit in the early years but also a low positive
cash flow from Years 3 to 5.

In the case of Development A, the amount of
money which has to be borrowed in the early years may
be too large for the bank to consider. Although the
medium-term prospects are good, the farmer may find
it hard to obtain finance. Even if he obtains a loan,
should markets or seasons be adverse during the first
or second year, the farmer may find that he has
underestimated his need for loans. He may then face
bankruptcy or, at least, severe financial difficulties.
With Development B, the risk is low, but so are the
profits in Years 3–5. Net cash flow from Year 5
onwards is similar to that for Development A. The
format for calculating cash flows is shown in
Worksheet 5, (p. 86).

The choice of which development to accept will
depend on the farmer's attitude to risk, his access to
credit, the value he attaches to an early rather than a
late pay-off, and the relative size of the positive and

Worksheet 10.1. *Partial budget for a simple change ($)*

Present activity 0.2 ha of tomatoes		Proposed activity 0.2 ha of fresh beans	
Income		*Income*	
Sale of tomatoes	400	Sale of beans	450
Grazing stubble	30	Hay from stubble	150
Total (*A*)	430	Total (*D*)	600
Running costs		*Running costs*	
Direct running costs (e.g. seed, boxes, sprays, fertiliser)	130	Direct running costs (e.g. seed, fertiliser etc.)	120
Total (*B*)	130	Total (*E*)	120
Annual profit from present activity (*A* − *B*)	300 (*C*)	Expected annual profit if change to new activity is made (*D* − *E*)	480 (*F*)[a]

[a] Advantage in favour of new activity (Beans) = $F - C$ (or $480 - 300$) = 180

negative cash flows of the two developments. Before a rational choice can be made between two developments, the farmer and his adviser need to know not only the return on marginal capital but also the pattern of pre-loan net cash flows.

Expected change in assets and debts

A new activity frequently requires borrowed funds. The increase in debts means higher interest charges and principal repayments. At the same time, the investment should raise the value of a farm. Spending $1000 on improving a farm is no guarantee that its market value will increase by $1000. Sometimes it can increase by more, sometimes by less. When a new development is being undertaken, it is usually worthwhile to calculate the expected change in value of assets and debts over the first 4–5 years of the development's life. There are two main reasons for this:

(i) to avoid the possibility that the level of debts will rise to too high a level in relation to the assets (if debts become too great, there is a danger of bankruptcy);

(ii) to assess the likely change in net worth (or equity) as a result of the project.

Depending on the conditions of ownership, increase in net asset worth and market value can be the most attractive feature of an improvement or intensification programme. The movement in assets and liabilities can be shown as in Worksheet 6 (p. 87).

Human and social aspects

Even though a planned change may look attractive on both economic and technical grounds, the expected result may never occur because of human and social obstacles. Lack of ingenuity and skill can be one reason for plans not succeeding. The customs and beliefs of the persons involved, their attitude to work, their sense of what is fair, their share of the benefits, their levels of expectation, may all make or break the new plan. So the farmer has to assess whether these factors are likely to cause trouble, what steps he can take to avoid it, and at what cost. Such human and social factors can be particularly important for the success of programmes which introduce changes within traditional tropical environments.

Intangibles

There will still be intangible factors, difficult to quantify, in arriving at a decision whether or not to go ahead with a change. They include a farmer's and his family's own preferences, values, and ambitions. A final decision is possible only after the intangibles have been taken into account.

An illustration using worksheets

The following budget worksheets are used to illustrate how to budget both simple and complex changes. Worksheet 10.1 covers a simple change – from tomatoes to beans, on 0.2 ha of the farm. If the crops were

Worksheet 10.2. *Partial budget for proposed complex change ($)*

Present activity 2 ha sorghum			Proposed activity 2 ha maize		
Annual income from crop and by-products	750		Annual income from crop and by-products	1250	
Total (A)	750		Total (D)	1250	
Annual running costs (not capital costs)			*Annual running costs (not capital costs)*		
Direct running costs (e.g. seed, fertiliser, wages, husbandry, repairs)	360		Direct running costs (e.g. seed, wages, sprays etc.)	500	
Depreciation only on items directly involved in the present activity	40		Depreciation only on items directly involved in this activity	50	
			Interest on any capital invested to make the change	60	
Total (B)	400		Total (E)	610	
Annual 'profit' from present situation or activity (A − B)	350	(C)	Expected annual 'profit' from new activity if the change is made (D − E)	640	(F)

equally risky, then a change to beans should be considered, as they also fix nitrogen. However, the physical and technical aspects of making the change need a good deal of investigation before going ahead.

Next (Worksheets 10.2–10.6) a more complex situation is described, where a small amount of capital investment is also needed to make the change. The worksheets are used to illustrate the expected financial effects of growing 2 ha of maize instead of the 2 ha of sorghum currently grown. It is assumed that the farmer already has some debts, which he is gradually paying off.

There are six worksheets. In the first two sheets are set out a partial budget (i.e. only the costs and income which would change if the farmer were to grow maize instead of sorghum are included). On the left-hand side of Worksheet 10.2 we show the annual profit (before deducting overhead costs) from the present activity, detailing the income and costs of growing sorghum. The right-hand side of the worksheet shows the figures for maize, including interest on any new capital needed for the maize activity, such as a planter. The figures for maize are those expected when production has become stabilised. When a new activity is planned that does not replace an existing one, no entry is made on the left-hand side of the worksheet.

Worksheet 10.3 shows the difference in 'profit' between the proposed new situation and the present. Account is taken of whether costs of income in other parts of the farm would alter if the change was made. Also, since extra profits may be taxed, allowance is made for this. The final figure shows the difference in profit after paying interest on the extra capital needed to bring about the change, and any taxes on the extra net income. In Worksheet 10.3, L is the net financial benefit if the change were made. If capital has to be invested in order to carry out the new activity, we can assess its earning rate by expressing L as a percentage of the net amount of capital invested. If the rate of return after interest and tax is less than 10% then the project is not very sound. The maize will bring a higher expected income than the sorghum, but its costs are higher. Even so, the expected increase in profit from the maize, after paying tax and interest on the extra capital needed is $306.

Worksheet 10.4 is a capital budget sheet which shows the net amount of extra capital needed to make the change. It also shows the expected percentage rate of earnings of the extra invested capital. The only investments needed are for a maize planter and special storage for the grain. No capital items will be sold. The expected increase in profit, expressed as a percentage of the extra capital invested is 51%, which is a very attractive use of capital.

Worksheet 10.5 shows the expected net cash flow for the whole farm if the change is made. It is necessary to illustrate the effect of the extra borrowing and income on the total farm finances because the farmer

Worksheet 10.3. *Appraisal of expected difference in annual operating profit if change is made ($)*

Expected difference in profit if change is made $(F-C)$: $(640-350) =$	290	(G)	Extra tax on net difference in annual profit	34	(K)
Has anything been overlooked?			Increase in annual 'profit' from change after paying interest on extra capital and tax on extra profit $(J-K)$: $(340-34) =$	306	(L)
Would costs or returns on any other parts of the farm alter if change to new activity is made? If so, what would be the net amount involved (H)	+50				
Net difference in annual 'profit' if change is made $(G+/-H)$: $(290+50) =$	340	(J)			

has other borrowings and there has been a loss of revenue from the sorghum. These items cannot be ignored when the farmer is making the decision whether or not to make the change. In those cases where no extra borrowings or capital investment are needed to make the change, there is no need to fill in the capital, cash flow and equity sheets. This view of the expected cash position of the total farm finances shows that the farmer would need to borrow only a small amount of extra money in the first year. He is still able to keep up the repayment on his long-term loan of $5000. In the later years he is able to reduce his debt.

Worksheet 10.6 shows the movement in the farmer's capital or equity (i.e. assets less debts) which is expected to occur over the next 3 years if the change is made. The data analysed in this worksheet shows that a gradual increase in equity can be expected, due to a slight increase in land value plus a reduction in debt (see pp. 86–7).

Note: these worksheets are used to illustrate the framework of the thinking behind, and the type of item which goes into, testing a proposed change. It is important, when planning a change, that the decision-maker uses these worksheets as a basis, but modifies them with factors and information from his own situation.

Decision-making and risk

Decision-making, in farming as in all business, involves choosing between possible alternative actions. It is one of the most important activities which a farmer has to do. It is also an act of management that cannot be ignored or postponed. Failure to take decisive action, when a choice is possible, is in itself a decision for which the consequences are just as real as those resulting from an overt decision and action by the decision-maker.

The key to decision-making is to approach problems in a systematic manner. The principal benefit to be gained from more formal approaches lies in the greater likelihood of making a good decision. This is because, with more formal techniques, all pertinent information is consciously brought together and a consistent and logical process of choosing is used to select the action or set of actions most likely to achieve the desired objective. Such procedures require clear objectives.

In briefly outlining this overall decision-making procedure, it is first necessary to distinguish a 'good' decision from a 'right' decision. A good decision is a rational decision. It assigns weight to each of the relevant factors involved in the decision. The weights assigned reflect the decision-maker's real beliefs founded on his experience and the information he was able to obtain about the alternatives facing him. A good decision is also consistent with the decision-maker's preference.

Since most decisions are made in the face of uncertainty, there is no guarantee that a 'good' decision will be the 'right' decision, i.e. whilst the decision-maker may act rationally, following formal decision-making procedures, he has no control over the outcome. For example, instead of selling his crop for $200 per tonne, a farmer may decide to store it with the prospect of getting $300 in 1 month's time. The best information he can get, combined with his experience, leads him to assign 90% chance (probability 0.9) of getting $300 per tonne in one month's time, but only a 10% chance (probability 0.1) of getting $140 per tonne.

Worksheet 10.4. *Capital aspects of proposed change ($)*

New capital expenditure involved in change			Annual interest on capital borrowed to make change		
Animals		0	Total capital $600 × interest rate 10% (from *E*, Worksheet 10.2)		60
Machinery – secondhand maize planter	500				
Structures – maize storage	100		Data to calculate rate of return on extra capital		
Other	0		Increase in 'profit' from change after paying interest, and tax on extra net income (item *L* from Worksheet 10.3)	306	(*E*)
Total (*A*)		600			
Capital items which would be sold if the change is made					
Animals		0	Net capital costs of change (*C*)	600	
Machines		0	Rate of return on extra capital used (*E* as percentage of *C* above)		
Other		0			
Total (*B*)		0	$\frac{306}{600} \times \frac{100}{1} = 51\%$ (*F*)[a]		
Net capital cost of making change (*A* − *B*)		600 (*C*)			

[a]The figure in (*F*) is after interest has been paid. The return on marginal capital before interest is 51%.

Worksheet 10.5. *Expected total farm net cash flow before borrowing if change is made ($)*

	Years					
	1	2	3	4	5	6
Cash receipts (excluding loans)						
Sales of all farm products	3300	3500	3600			
Other receipts (not loans)	200	200	200			
Total (*A*)	3500	3700	3800			
Cash payments						
Variable costs	1200	1300	1350			
Overhead costs	600	700	750			
Loan repayments	500	500	500			
Capital investment	600					
Interest on previous debts	200	100	50			
Living costs	600	700	800			
Taxes	200	300	300			
Total (*B*)	3900	3600	3750			
Net cash flow before borrowing (*A* − *B*)	−400	+100	+50			
Annual borrowing needed to meet payment deficiency	400	—	—			

Due to some unusual factor in the market, the actual price received is $140. In this case, the 'good' decision has a bad outcome, so it was a 'wrong' decision.

Attitudes towards risk

Farm structure is an important factor affecting an individual's attitude to risk. Many farmers in Africa, Asia and Latin America operate so small a unit that, even in good years, they have little margin beyond their essential needs. Often managers of large farms which are organised as cooperatives are reluctant to take risks because they are responsible for a large number of members. State farms are sometimes used to try out new ventures as the government can better afford to take the risk.

Probabilities

An understanding of how to use the concept of probability can help in decision-making. By estimating probabilities a farmer is able to apply some measure to the risks he takes.

When a coin is thrown into the air the chance, or probability, of it landing head upwards is 1 out of 2, i.e. 50% (or 0.5, as it is commonly written). The likelihood of the coin falling tail upwards is also 1 out of 2 or 0.5.

Worksheet 10.6. *Expected movement in value of assets and debts ($)*

	Years				
	1	2	3	4	5
Assets					
Land and improvements	15 000	15 500	17 700		
Machinery	1500	1300	1400		
Livestock	3000	3000	1300		
Bank					
Total (*A*)	19 500	19 800	20 400		
Debts					
Short-term loans	1000	1200	1000		
Long-term loans	5000	4500	4300		
Sundry creditors	500	500	500		
Total (*B*)	6500	6200	5800		
Equity or net worth (*C*) (*A* − *B*)	13 000	13 600	14 600		
Equity % = $\frac{C}{A} \times \frac{100}{1}$	66.6	68.6	71.5		
Annual debt service cost					
Interest	200	100	50		
Loan repayment	500	500	400		
Total	700	600	450		

Table 10.3. *Chances out of 36 (i.e. 6^2) of 2 six sided dice coming face up with a specified total*

Total	Chance out of 36	Chance out of 100	Approximate probability
2	1	2.7	0.03
3	2	5.4	0.05
4	3	8.3	0.08
5	4	11.1	0.11
6	5	13.5	0.14
7	6	16.4	0.17
8	5	13.5	0.14
9	4	11.1	0.11
10	3	8.3	0.08
11	2	5.4	0.05
12	1	2.7	0.03

The frequency with which something happens or occurs can be expressed as a fraction of the number of times it could have occurred. This (and any other) information can be used by the decision-maker to form his own estimates of the likelihood or probability of something happening. If a farmer knows that during the past 20 seasons he has had a crop yield of 1 tonne/ha for 14 seasons, 0.5 tonne/ha for 4 seasons, and nothing at all for 2 seasons, and if he cannot see any significant change for the next (21st) season, then he might expect that there is 0.7 probability of getting 1 tonne/ha (14/20); 0.2 probability of getting 0.5 tonne/ha (4/20); and 0.1 probability of having a total crop failure (2/20).

A more complex example of this idea is the throwing of two six-sided dice. The chance (probability) of obtaining different totals by adding the numbers shown on each of the two dice after throwing are as follows (see Table 10.3). The extreme totals (1 or 12) are least probable: totals from 5 to 9 are more likely (more probable).

Assessing risks in terms of the probability of various outcomes is a help in production planning. Thus a farmer who sees possibilities in growing vegetables on the flood plain of a river, would find out from his neighbours or his adviser how many times within memory the river had flooded and caused the loss of a crop.

If it appeared that flooding occurred, on the average, once in every 5 years, the risk factor (20% chance of crop failure due to flood) would become a basic element in his production plan. If the profit from cultivating for 4 years is sufficient to cover the consequences of a total loss of all his outlays one year and still leave an attractive margin, then production would be worthwhile. The risk of flooding would then have been recognised, appraised, and allowed for in the production plan.

Variability in rainfall can be considered in the same way. If the rainfall in a particular area is sufficient and timely enough to support a good crop in only 3 years out of 5, the returns from these 3 years must be sufficient to cover the losses on the other two, and still leave a profit. Otherwise it would be unwise to start cultivation in this area without access to supplementary irrigation.

The problem is more complicated where there is a close interaction between a major input and a variable influence such as rainfall. If fertiliser is applied generously and rainfall is inadequate, then the farmer loses money on the fertiliser applied. On the other hand, if he does not apply the optimum amount of fertiliser for fear of lack of rainfall, then if there is adequate rain, he misses out on the possibility of a much larger yield and a higher return. The best thing to do would be to apply that amount of fertiliser which would maximise returns at the level of rainfall which is most probable.

One difficulty in using probabilities to help decide how much extra input to use is lack of knowledge, for example, of the frequency distribution of rainfall. Without any other sources of information farmers follow tradition and their own experience, which will obviously be coloured by recent events. Another difficulty is that many farmers have too little capital to take the risk of applying fertiliser when there is a chance that inadequate rainfall will mean it is applied at a loss in some years. For example, it is necessary to apply extra fertiliser to varieties of rice developed by the International Rice Research Institution (IRRI), in order to get the full benefit of its growth and yield potential. This need delayed the adoption of IRRI rice in some parts of India where dry seasons or inadequate irrigation facilities could result in negative returns to outlays on fertiliser.

Taking account of variability

So far, in our budgeting examples we have used single values for costs, yields and prices, e.g. annual costs $500, annual income of $1000, yield 1000 kg/ha. In practice, there is often a range of possible outcomes. Any farmer who does not take account of the likely variability will be inadequately prepared if the actual outcome is different from the expected result.

Because of the variable nature of the weather and agricultural markets, no farm plan is likely to work out exactly as expected. Therefore, it is necessary to state what would happen if prices and yields were (i) less favourable and (ii) more favourable, than expected. The percentage chances (probability) of these events happening should also be specified. Worksheet 10.7 (p. 000) shows the procedure, based on the example earlier in this chapter (p. 000). In Worksheet 10.7 we show how the extra profit from the switch to maize is expected to vary. The effect of a crop failure on the total farm finances, and the effect of an especially high yield, are also demonstrated.

Worksheet 10.7. *Expected variation in increase of profit if the change is made*

	($)	Chance (%)
Outcome A: 'Most likely' profit increase	306	70
Outcome B: 'Pessimistic' profit increase	− 300	10
Outcome C: 'Optimistic' profit increase	500	20

Effect on Farm Finances

	Outcome B	Outcome C
Serious	If in first or second year	
Not substantial	If occurs after first 2 years	
Good		Yes, especially if in early years

Plan of action if outcome B: do not add to house
Plan of action if outcome C: reduce debts; repair house

If pessimistic outcome:
How seriously will this affect the total finances of the farm?

(Very seriously/seriously/not seriously)

If optimistic outcome:
What effect will this have on the total finances of the farm?

(Great benefit/moderate benefit/little benefit)

There is a strong chance that the switch to maize will be profitable but there is also the possibility of the farmer making a loss. His decision to go ahead with the programme will depend on the state of his total finances, and on his attitude to risk. In this technique we look at the total picture and incorporate various probabilities. A budget using 'most likely' single values is prepared and the expected annual profits or the annual net cash flows are calculated. Then the possible variations around the 'most likely' figure are estimated, together with their percentage chance of happening. The farmer then has to specify what action he has planned in the event of an outcome, other than the 'most likely' one, occurring.

Annual operating profit

It is sometimes useful to draw up an annual operating profit budget. The likely variability of the profit should then be assessed. This estimate of variability can then provide a basis for working out contingency plans should the 'worse than' or 'better than' expected situation happen (as we did in the example above).

Expected money values

Expected money values is a concept used to take some account of variability of key factors such as yield or prices, when doing budgets. Suppose that the outcome, in this case net 'profit', from a hectare of crop in a particular area could vary as follows:

Event	Outcome Net profit/ha ($)
Bad year	− 30 (loss)
Poor year	0
Reasonable year	+ 36
Good year	+ 84

The farmer and his adviser then decide, on the basis of their experience that in, say, 20 years, they could expect a 'reasonable' year in 10 years, 4 years to be 'poor', 4 years to be 'good' and 2 years to rate as 'bad'. This gives probabilities of the four types of performance as:

Event	Frequency during 20 years	Probability[a]
Bad year	2	0.1
Poor year	4	0.2
Reasonable year	10	0.5
Good year	4	0.2

[a] Adds up to 1.00.

To get the expected money value (EMV) of each event, the outcome – in this case 'profit' – is multiplied by the probability that it will occur. Thus if it were a bad year, the EMV per hectare would be ($-$30) × 0.1 (probability) = $-$$3. In Table 10.4 below, we show how to calculate the total EMV for the hectare of crop.

We think that it is better to use EMVs when planning than it is to use crude averages. At least, it forces the decision-maker/adviser to recognise that things can go better or worse than the 'average'. Certainly, there are problems associated with using

Table 10.4. *Total expected money value for one hectare of crop*

Event 1	Outcome ($) 2	Probability of event happening 3	Expected money value of outcome 4 (2 × 3) ($)
Bad year	− 30 (loss)	0.1	− 3.00
Poor year	0	0.2	0.00
Reasonable year	+ 36	0.5	+ 18.00
Good year	+ 84	0.2	+ 16.80
Total expected money value			+ 31.80[a]

[a] The $31.80 is the probability weighted sum to be used in budgets.

what is termed, in economist's jargon, 'subjective probabilities' (measures of personal strengths of belief about uncertain events happening, measured on a scale from 0 to 1). Since it is the farmer who has to make the decision, and to live with the consequences of that decision, we don't know of any better way of giving him a sensible basis for making up his mind. Furthermore, since the farmer has first-hand knowledge and experience of the environment in which he has to operate, he will have the best idea of what the likely outcomes are. He is acting rationally if, in the light of all the information available to him, he assigns probabilities to a range of uncertain future events for the years ahead.

The outcomes of each event, and the associated probabilities, need to be taken into account before a final decision to grow a crop is taken. If the farmer is unable to feed his family properly should a bad or poor year occur – a total probability of 0.3 (or 3 chances in 10) he may choose to grow a crop which has a lower probability of failure, even though its expected money value may be only $20/ha. If it did not make much difference to him if the 1 ha of crop grew badly or poorly, and if he were also a bit of a risk taker, then he would probably choose the crop with the highest EMV, because there is a 40% chance of him getting a reasonable profit at $36, and a 20% chance that he would get a good profit ($84).

Break-even and parametric budgeting

One way to take account of the effect of variability is to calculate what the minimum price and/or yield and/or gross income per hectare would have to be before the proposed change in activity produced only the same amount of profit as the one presently being conducted. This technique is called 'break-even budgeting'. A more detailed method is to vary both expected prices and yields over a range of possibilities, and to calculate the change in 'profit', after interest and tax, for each of these (parametric budgeting).

As we have already observed, a budget is no more reliable than the data used to construct it, and some (often much) of this data may be uncertain. While the technique of break-even and parametric budgeting, described below, do not actually solve the problem of having uncertain data, they do provide the farm operator with an indication of what will happen if prices, costs or rates of performance vary from the values assumed in an ordinary budget.

Break-even budgeting

Break-even budgeting can best be used when one particular component of a partial budget is uncertain, and the chances that the proposed change will prove to be profitable have to be assessed.

The method is applied by representing the uncertain component (e.g. price of maize) with a symbol, say X, and then constructing the budget in the usual way. The increase or reduction in profit is assumed to be zero. That is, it is assumed that the proposed new activity just breaks even with the existing activity, so that the totals of the two sides of the budget are equal. Once the break-even value for the uncertain part has been found, the chances that the actual value of the component will be either better or worse than this break-even value can be assessed.

Suppose that in the example discussed earlier – changing from sorghum to maize – the farmer felt uncertain about the price of maize. The farm adviser could calculate the price per kilogram required to break even, as follows.

An example of a break-even budget when changing from sorghum to maize

We will assume that in the partial budget (Worksheet 10.2, p.84) the $1250 income was obtained from 10 000 kg sold at 12.5 ¢/kg. We will calculate the price/kg maize has to drop to before the extra profit

resulting from the change to maize drops to zero. Put another way – with the proposed new maize activity, what is the price of maize which makes the profit 'break-even' with (be equal to) the profit from the existing crop, sorghum? From Worksheet 10.3 (p.85) the extra 'profit' from the change from sorghum to maize is $306. Let the break even price of maize be X cents/kg, with 10 000 kg of maize grown.

X is the price of maize at which the farmer is no better off than he was before the change. The total benefit from having the maize is equal to the maize gross income, $(10\,000\,X)$ plus the $50 due to other benefits (item H in Worksheet 3). Thus 10 000 kg of maize has to sell for price X in order to break even with the current profit from the present crop (sorghum).

The $10\,000\,\text{kg} \times X$ plus the $50 of other benefits have to equal (break even with) the sum of the present profit of $350 from sorghum, the $610 costs of growing the maize, and the $34 paid in tax. The way to calculate the value of X is shown below. The data were derived from Worksheets 10.2 and 10.3. The break-even equation is:

Gross income $(10\,000\ X) + \$50$ (other benefits)
$= \$610$ (maize cost) $+ \$350$ (sorghum profit) $+ \$34$ (tax),
$$10\,000\ X + 50 = 610 + 350 + 34,$$
$$10\,000\ X = 944,$$
$$X = \frac{944}{10\,000} = \$0.0944 \text{ or } 9.44 \text{ ¢.}$$

Should the price of maize exceed 9.44 ¢/kg the break-even point is passed. If the farmer–decision-maker thinks it most unlikely that the price of maize could fall to this level, and if no other options exist, then he has some reason to go ahead and grow maize instead of sorghum.

The other main use of break-even budgeting is if the price is known with certainty, e.g. a contract price which has been agreed to by a buyer. In this case the break-even yield is worked out, using the reasoning in the above example.

Parametric Budgeting

Parametric (or flexible) budgeting is a budgeting technique in which a range of prices and yields are considered at once. A parameter is any factor which has an influence on the net profit of an activity or project. Parametric budgeting involves a series of budgets, expressed as equations, in which the parameters vary. Usually, in practice the number of important parameters which do vary is small, thus limiting the amount of calculation necessary. Yield, price per kilogram, and variable cost per hectare are key parameters in a cropping activity. The steps in constructing a flexible or parametric budget are:

identify the key parameters in the activity or project;

construct a profit equation by expressing the relationship of the parameters in simple mathematical form;

solve the equation using different parameter values and express the results either in figures or as a graph.

The following examples illustrate parametric budgeting. For maize production the profit equation can be written as follows:

Profit per hectare $= Y \times P - V$,

(where $Y =$ yield, $P =$ price, $V =$ variable costs). If some key variable cost is directly related to the amount of yield, such as harvesting costs (H), then the equation is:

Profit per hectare $= Y \times (P - H) - V$.

If fertiliser is an important parameter which is likely to vary in the amount applied, and in cost, then it can be identified separately. For example:

Profit per hectare $= Y \times P - (F + V_o)$,

F means fertiliser costs, and V_o means variable costs other than fertiliser. If the yield of maize is 5000 kg, the price 10 ¢/kg fertiliser cost \$60, and other variable costs (including marketing) \$200/ha, the profit equation would be:

$$
\begin{aligned}
\text{Profit per hectare} &= 5000 \times 0.10 - (60 + 200) \\
&= 500 - 260 \\
&= \$240.
\end{aligned}
$$

Different values of the key parameters can be substituted in the profit equation, thereby a range of problems account for a range of possible yields, prices and costs can be calculated.

Sensitivity analysis

In many activities or projects, the net profit depends on the values of just two or three key parameters. Sensitivity analysis, which is normally based on a parametric budget approach, shows:

those parameters (yield, prices or costs), which have the greatest effect on net profit;

the extent to which the size of the net profit is sensitive to a change in the value of one or more of these parameters.

For example, in a given maize production activity, the size of the net profit or gross margin (GM) may depend mainly on the yield (Y) and the cost of fertiliser (F) used. The net price (P_n) after deducting harvesting, marketing and pre-harvest variable costs, other than fertilizer (V_o), may be fairly certain. Then the relationship can be expressed in the profit equation below:

GM per hectare $= (Y \times P_n) - V_o - F$.

If the yield is fairly closely related to the amount (and hence the cost) of fertiliser used, the extension worker can substitute likely values of Y and F in the equation. This tests the sensitivity of the gross margin to changes in these values. The range of gross margins that result from the sensitivity test will give him and the farmer a basis for deciding on the amount of fertiliser he will need to use in order to get the 'most likely' highest gross margin. The information from the sensitivity test will further stimulate the farmer to focus attention on the need to use the proper crop-production techniques for obtaining the expected result. Thus he is likely to be more conscious of the need to take timely and appropriate action if, say, weeds threaten his crop. Also, he will be on the lookout for early signs of insect pests or diseases which could cause the yield to fall.

Sensitivity analysis can also be applied to the total net profits of projects, and here it serves three functions:

(i) indicating the range of possible financial outcomes;

(ii) emphasising to the farmer the need to use the proper crop-production techniques and the need to be flexible and timely in his response to both adverse and favourable developments;

(iii) stimulating him and his adviser to think about, and to plan, courses of action in the event of the outcome being other than that deemed to be the 'most likely'.

Climatic and yield variability

There are several ways of reducing the harmful financial and physical effects of uncertainty about the future. One is to keep the equity high (i.e. do not over-borrow, but keep the equity in the farm high to allow access to further funds for, say, replanting). In general, the equity percentage should not drop below 70% even in the less risky areas. This 'rule of thumb' figure is, of course, a rough guide because the level of net surplus disposable income, and living expenses and tax levels will also affect the safe lower limit of equity.

Another method is to build into each yearly budget a contingency sum to cover items of unexpected cost or reduced income; in years of good profit, invest only some of the surplus in the farm. It is often a good idea to invest the rest in safe off-farm investment in areas which have no relation to agriculture. That is, where the investment can be sold easily (jewellery), or shows a sound prospect of capital growth (increase in value) or is able to earn some income (e.g. small store or a minibus taxi).

On crop farms, fluctuation of income due to variability of the climate can be reduced by diversification, and by having the tools, plant, technical knowledge, access to working capital and labour to permit a quick and flexible response in land use to make the most of seasonal opportunities as they occur.

Where there is a chance of a total crop failure, say once every 5 years, if possible an allowance of one fifth of the total pre-harvest variable crop costs can be added to each year's costs, and the money put into some interest-earning investment which can be readily converted to cash. If this precaution is taken there will be, in general, funds available to meet pre-harvest costs in the year following one of crop failure.

When a crop activity fails during the early or middle part of the year it is often still possible to follow it quickly with another, thereby compensating the loss of the first activity. If the farmer is not prepared, in terms

of knowing the technical requirements of the next or alternative activity, and having quick access to the inputs (seed, labour) needed, then it is likely that he would not be in a position to make a successful, opportunistic, change of plan.

For the semi-subsistence farmer, the major ways of avoiding financial difficulties due to variability are:

diversification of activities (where it will ensure a more reliable food supply);

storage (where produce can be protected against deterioration, pests and predators);

keeping savings in a readily convertible form.

Some specific measures to cope with income variability

The techniques discussed earlier are used mainly to indicate the likely effect of variability and to stimulate the farmer and his adviser to make contingency plans. Now, we will discuss some of the measures available to help farmers reduce income variability (i.e. to keep net profits at a fairly stable level), in the face of price and climatic variability. Means of reducing income variance include:

Yield insurance

Hail, flood, fire and sometimes frost insurance can be obtained for some crops in a number of countries. Livestock are more difficult to insure. The premium for insurance is in proportion to the risk borne by the insurance company.

Adoption of sound crop- and animal-production techniques

A common explanation of why one farmer makes a success of an activity in a given season, while his neighbour makes a loss, is that the first farmer paid proper attention to the basic husbandry techniques needed to produce the crop or animal product. For example, good fallowing and weeding techniques, getting operations done on time, proper depth and spacing of seeds, timely use of correct sprays and fertiliser, preparation of a good seed-bed all give a better chance of success, no matter how the season turns out. The use of good methods and techniques as a means of reducing the effects of risk is often not given enough attention.

Growing crops for which there is a guaranteed price

For some commodities, government marketing boards or cooperatives guarantee the minimum prices of some products. The farmer has the choice of growing these products and thus reducing the variation in his income. In some years, however, his income would be higher if he grew products for which there was no fixed price.

Diversification

Diversification can frequently be of value when there is variability due to price or weather conditions. On the other hand, specialisation has obvious advantages, since resources (including accumulated managerial skills) can be devoted to producing the product which will give the highest return. Where the relationship between two activities is competitive for a scarce resource, including the less profitable activity in the farm plan is usually not warranted (unless for biological reasons, such as in crop rotations).

Diversification allows flexibility of resource use and can lower income variability. Sometimes the returns from different enterprises are affected by the same factors, such as weather or market conditions. In such cases, a reduction in the gross margin of one enterprise is matched by a similar reduction in the gross margin in another enterprise. This diversification into a number of 'similar' enterprises can result in an increase in the amount by which income varies.

In areas of great production uncertainty, little benefit in terms of reduced income variance can be obtained by diversification. As production certainty increases there is an increase in the benefits from diversification into activities whose GMs show least correlation, e.g. poultry and beef.

Contracting to processors or merchants

The processor or merchant guarantees a fixed price for a specified amount of product and/or to take a given quantity. This arrangement is common with tobacco, sugar cane, poultry products, fruit and vegetables for processing.

Hedging on future markets

Where such markets are in operation a farmer can guarantee himself a known price now, rather than wait for an uncertain price 3–4 months hence. Futures markets are not common in tropical developing countries.

Keeping borrowings to a level where they can be serviced

The ability to service debt depends on the level of the annual cash surplus which results from the year's operations before deducting interest and principal payments. It is calculated as follows:

(A) Gross cash income
   Less cash operating costs
   (including living)
       Net plant replacement costs
       New capital investment
       from 'profits'
       (not from borrowing)
       Taxes and other cash
       outgoings

(B) Total cash outgoing before debt
   servicing

(A–B) Annual cash surplus (available
   for interest and loan repayment)

In calculating the annual cash surplus, depreciation is omitted. The higher the annual cash surplus, the higher the debt that can be serviced and the lower the critical equity percentage necessary for the farm business to survive.

$$\text{Equity } \% = \frac{\text{Assets} - \text{liabilities}}{\text{assets}} \times \frac{100}{1}$$

Whenever undertaking a programme which involves borrowing, the farmer and his adviser need to consider how much money would be left to service debts (i.e. pay off interest and loans), if the programme did not work out as expected. They also need to consider what would happen if the money was not enough to service the debts. Would his banker lend more money for next year? If yes, could he meet the extra interest and repayment burden at the end of next year? What is the probability of him being able to manage this? These questions need to be answered before deciding on how much money to borrow.

Use insurance if available

Heavy financial losses can be caused if assets are destroyed or stolen. Also, if accidents occur, the farmer can be liable for a large payment to the injured person. The amount of money involved can sometimes be enough to bankrupt him. An insurance policy can protect the farmer against these eventualities.

Flexibility

The ability to be able to change plans and activities quickly is important in reducing income variability. It involves planning ahead so that the means of making a quick response are available to the farmer.

Questions

1 Give an example from one of the local farms of a change which would be (i) simple, and (ii) complex.

2 'Budgeting tools are just practical ways of applying economic principles to decision making'. What do we mean by this?

3 Would 10–15% return on marginal capital be enough to allow for the riskiness of adopting some big changes to farm plans in your area, or would at least 30% be needed before the local people would think it worth trying it out?

4 The worksheets in this chapter are just examples. Suggest how the methods of evaluating changes which we describe could be made more useful to you. (We are offering $A10 for anyone who comes up with a good idea, and we will buy them a few beers as well!)

5 I'll gamble on cards, dominoes, and races, but I can't afford to take too many risks with growing food for the family, can I? Can you put odds on the chances of yields and price levels occurring, in the same way as is done in other games of chance?

6 If you tossed a coin six times, and each time it came down heads, then it just has to be tails next time, doesn't it? Do you agree? Does the coin remember?

7 Last season was bad. Next season has to be good, according to the law of averages. Is this true?

8 It doesn't matter how often something has happened in the past, what is important in decision-making is how strongly you believe something will happen in the future. What do you understand by the term 'probability'? How can you use probabilities in farm management planning?

9 Explain what EMV means.

10 Break-even budgeting gives you more information to work on than budgeting with single 'best guess' figures. How?

11

Cropping

Introduction

In all cropping activities there are four main operations:

(i)   preparing the seed-bed;
(ii) growing (planting, seeding, watering, weeding, fertilising, spraying, protecting, pruning);
(iii) harvesting;
(iv) marketing (which can include processing, sorting, packing and transport).

Also, there are some production, harvesting and marketing decisions which are common to all crop-production activities. These include:

which crops, and combination of crops?

which variety of crops?

what husbandry practice for each crop?

what cropping system should be followed (sole, rotation or intercropping)?

how to maintain soil fertility?

how best to ensure timeliness of operation of such practices as sowing, weeding, disease control, spraying and harvesting?

how to use labour and equipment for maximum efficiency?

where and when to get casual labour?

how to obtain the best technology to get most profits in intensive, risky and disease-prone production systems?

where and how to sell the produce?

what to do with the produce if prices are low (frequently storage is not possible and losses from insects and pests – including two-legged ones – can be high)?

what will be the likely price at harvest?

whether to plant all the crop at the same time, or to stagger plantings?

Generally, four types of cropping activities can be recognised. These occur when the crop

(i)   grows quickly (3–4 months) from planting to harvest, e.g. many soft vegetables such as tomatoes or beans. Often, two or more such crops can be grown on the same area of land within 1 year.
(ii) takes 4–7 months to grow – most of the commonly cultivated grain crops (such as maize, sorghum and millet) are in this group.
(iii) has about a 12 month growing season (yams and cassava are examples).
(iv) takes 2–3 years to produce the first harvest, maybe 5–6 years after that to reach full production, and continues to produce for 20 or 30 years. Tree crops such as fruits, oil palms and nut trees belong here.

We will discuss crops in four sections – rotations, mixed cropping, harvesting and marketing, and tree crops.

Rotations

The main management objective of most farm families is to produce enough of the various crops which can be safely grown, to give an adequate, varied and palatable year-round diet for the family. Labour comes mostly from family sources, and there are times in the cropping cycle when family labour is inadequate for doing all the crop-husbandry jobs properly. This puts a limit on the area which each family can work, stimulates the farmer to use a cropping system which both reduces peak labour demands and spreads the work more evenly over the growing season.

The term 'rotation' can cover a whole range of cropping activities and sequences: multiple cropping, shifting cultivation, slash and burn, mixed cropping, intercropping, relay planting, and so on. Some important terms and possible systems of cropping are listed, with their common meaning, below (from R. R. Harwood in *Small Farm Development*, published by Westview Press, Colorado in 1979).

Cropping pattern The yearly sequence and spatial arrangement of crops or of crops and fallow on a given area.

Cropping index Number of years of cropping on a given field multiplied by 100, divided by the length of the rotation. This is called the '*R*' value (see p. 96 for a fuller explanation).

Multiple cropping Growing more than one crop on the same land during 1 year. Within this concept there are many possible patterns of crop arrangement in space and time.

Sequential cropping One crop is planted after the harvest of the first.

Monoculture planting Growing a single crop on the land at one time. Another definition is 'the repetitive growing of the same crop on the same land'.

Ratoon cropping The cultivation of regrowth from stubble following a harvest, not necessarily for grain.

Double cropping Growing two crops in the same year in sequence, seeding or transplanting one after the harvest of the other (same concept for triple cropping).

Strip cropping Growing two or more crops in different strips across the field wide enough for independent cultivation. The strips are wide enough to give greater intracrop than intercrop association.

Interplanting All types of planting of one crop into another crop which is already growing. It is used especially for annual crops grown under stands of perennial crops.

Intercropping A specific form of interplanting. Two or more crops grown simultaneously in the same, alternate, or paired rows in the same areas.

Interculture Arable crops grown below perennial crops.

Mixed cropping Two or more crops are grown simultaneously in the same field at the same time, but not in row arrangement.

Relay cropping (relay planting) The maturing annual crop is interplanted with seedlings or seeds of the following crop. If the flowering period of the first crop overlaps with the second crop in the field, the combination becomes intercropping.

By 'cropping system' we mean the particular mixture of crops, times of planting, weeding and harvesting for each, and the rotation followed.

Traditional methods of maintaining soil fertility have had to change in response to demands for more intensive land use. Thus, the bush fallow system which had 2–3 years of crop, followed by 7–8 years of bush fallow to restore fertility is going out of practice. In its place, there may be 3 years of crop, followed by only 2–3 years of fallow. In most cases, this fallow period is too short to allow fertility to be restored after the crop phase, so extra nutrients have to be added to the soil from purchased mineral and/or organic fertilisers. There is a measure of cropping intensity used by agronomists. They use the symbol R, and define it as the number of years of cropping, multiplied by 100, and divided by the length of the rotation. Thus if there were 3 years of cropping followed by 7 years of bush fallow on the same piece of land, there is a 10 year crop rotation. The intensity of cropping, R, is then

$$\frac{3 \times 100}{10} = 30.$$

3 years crops, followed by 3 years of fallow has an R value of 50, i.e.

$$\frac{3 \times 100}{6} = 50.$$

With 2 years cropping, followed by 1 year of fallow, R is

$$\frac{2 \times 100}{3} = 66.$$

The more the land is under crop during any rotation, the higher the *R* value. Cultivation systems can be classified on their *R* values:

Permanent	66 or more
Natural fallow	
(*a*) shifting systems	33 or less
(*b*) 'fallow' systems	between 33 and 66

Note: In intensive systems, '*R*' can exceed 100.

'Ley' systems do not use the bush, natural grass or savanna to restore fertility – they are based on sown, legume-based pastures which are grazed by animals during the 'resting' period between crops. Legume crops such as ground nuts (peanuts) can also be used in this role, but the amount of nitrogen they contribute to fertility is less if the crop is harvested, i.e. some of the nitrogen formed by the plant is removed from the area when the crop is harvested. A rotation of 3 years of crop followed by 4 years of ley (or ley crop) in a 7 year rotation has an *R* value of 43.

On semi-subsistence farms, another important aim of management is to reduce the chance that the supply and/or variety of food will not be enough to enable the requirements of the family to be met. To reduce the risk of this happening, farmers use a variety of techniques, the main ones being sowing 'traditional' rather than 'new' varieties and packages, staggered plantings, intercropping, moisture conservation before sowing (by removing weeds and ridging the soil), and timely weeding. In relation to the demand for land, because of population pressures, there is not an abundance of fertile land simply waiting to be cultivated in most developing countries.

In fact, many of the soils being cropped are low in nutrients, have many weeds and a poor structure. This is due either to their natural infertility or to exhaustive cropping over a long period. So the need for melding the traditional methods of maintaining soil fertility and modern fertiliser, weed control, cultivation methods and plant genetic technology, is becoming increasingly urgent.

Many cropping systems being practised are plundering the soil, not improving or maintaining it. In Hans Ruthenberg's classic text, *Farming Systems in the Tropics* (3rd edition, Clarendon Press, 1980), he comments:

The basic principle of farming is to change the natural system into one which produces more of the goods desired by man. The man-made system is an artificial construction which requires continuous economic inputs obtained from the environment to maintain its output level. Farming thus implies the abolition of an unproductive 'steady state' in favour of a man-created, more productive but stable 'state', and much of the farm input (tillage, fertilizers, weeding etc) is nothing but an effort to prevent the new state from declining towards an unproductive low-level steady state . . .

We may hypothesize from past events that economic development apparently passes through the stage of soil mining in the early phases of industrialization to the stage of 'soil improvement' in a highly industrialized environment. *This book is nevertheless biased in favour of balanced and improving systems* [present authors' emphasis]. At the same time, it is recognized that soil mining may be inevitable in economic development, and to ask for soil fertility maintenance as a *conditio sine qua non* [an essential precondition] in economic development is to ask for the impossible because the requirements in terms of inputs, and the short-term outputs foregone, are simply too high. The extreme conservationist tends to waste funds which are badly needed for more productive forms of capital development, and he may be as dangerous in terms of a stable world as the land miner. *The disquieting aspect, in my opinion, is not that soil mining occurs, but the world-wide extent of the phenomenon and the speed of the mining process in most of the tropics* [present authors' emphasis] . . .

Farm maintenance and improvement consequently takes place at the expense of stocks in the environment. Nutrients can only be accumulated on a given field at the expense of other areas or stores. The income-creating capacity of farming is thus a function of the effectiveness of the transfer of resources from places where they are useless or of low effect in terms of human objectives, to those where they serve man's goals, e.g. phosphate deposits.

Ruthenberg's comprehensive text is an excellent basis of technical knowledge upon which to build and apply economic principles and farm management techniques.

Factors affecting crop yield

There is usually more variation in income with cropping than with animals, as many of the factors which determine crop yield are completely outside the farmer's control. The weather, especially rain, has a far more important influence on yield than any other factor, i.e. very favourable weather can double the expected crop yield, whereas it will not double annual meat, milk or animal fibre yield. In Table 11.1 are the main variables in the cropping enterprise which can be

Table 11.1. *The main variables affecting crop yields*

A Large degree of control	B Moderate degree of control	C Little or no control
Species	Moisture stored in seed-bed (with fallow)	Time of sowing other crops
Variety	Soil structure	Frost and hail
Plant population (seeding rate and row spacing)	Some diseases	Moisture at harvest
Fertiliser	Time of sowing (some crops)	Moisture during growth (except irrigated crops)
Weeds	Many insects	
Some diseases	Length of sowing period	Wind
	Date of harvesting	Flood
Quality of seed-bed		Large-scale plagues of insects (e.g. locusts).

manipulated, to a greater or lesser extent, by the farmer in his efforts to increase profits. Column A contains those variables over which the farmer can have complete control. Column B has those factors over which he has varying degrees of control, while those which are virtually beyond managerial control are shown in Column C.

Depending on the crop, there are also varying degrees of uncertainty about price. Two big problems facing a crop grower are to decide the weight to attach to each of the variables lists above, i.e. the extent to which each affects yield; and the 'price' to pay to help counter the yield-depressing ones and to foster the yield-increasing ones.

Fertility

Continuous cropping of the same area of land normally leads to a decline in crop yield through loss of soil structure, depletion of nutrients (especially nitrogen), an increase in weeds and often a build-up of harmful fungi, bacteria, microorganisms and insects. It can also often cause erosion. What do we mean by soil 'fertility'? A soil is fertile, by our definition, if it has the following features:

adequate mineral nutrients;

a structure which encourages vigorous root growth and which retains moisture;

absence of weeds;

low levels of harmful microorganisms such as viruses, bacteria, fungi, and harmful soil animals (nematodes and insect larvae);

a large population of useful microorganisms and soil animals, such as nitrogen-fixing bacteria and earthworms (respectively);

little or no erosion.

One way of maintaining soil fertility on crop farms is to rotate a legume crop, bush fallow or native grass fallow phase with the crop phase. A period of 3–4 years of a vigorous legume phase or a longer period of, say, bush fallow improves soil 'fertility' (as defined), i.e. it improves nitrogen levels, soil structure, suppresses weeds, reduces harmful microorganisms and renders the soil less prone to erosion. The amount of nitrogen fixed by a legume varies with soil type, the vigour of the legume crop, the existing level of soil nitrogen, and the efficiency of nodule-forming bacteria.

The practice of alternating or rotating legume crops of pastures with grain or root crops is called 'ley' farming. Where the non-crop phase has bush or savanna vegetation, it is called bush or savanna fallow. Some of the main advantages of rotations with ley or fallow are:

building or maintaining soil fertility;

forage production;

better crop yields;

better soil erosion control;

improved weed control;

balanced depletion and replacement of nutrients;

the consistent use of a particular cropping system;

reduction in levels of harmful microorganisms.

Non-Fallow Systems

Fertility can also be maintained or built-up in cropping systems by chemical and mechanical means. In a number of areas, a *stable system which contains no legume or fallow in the rotations* has been developed by using:

fertilisers (especially nitrogen) to maintain the nutrient status;

Table 11.2. *Role of fallow/legume alternatives*

Influence on the soil of continuous cropping	Mechanical and chemical means which can help off-set the influence of continuous cropping
Depleting soil structures and increasing erosion	Special stubble mulching practices, anti-runoff contour banks, strip farming
Decreasing the amount of plant nutrients	Chemical (or organic) fertilisers
Decrease of organic matter	Retaining stubble and applying extra nitrogen
Increase in disease-causing organisms	Chemical pesticides, fungicides and bactericides

stubble retention to improve the structure and reduce erosion;

sprays, cultivation and alternation of winter and summer crops to reduce weed competition;

chemicals and rotation of cash crops to prevent harmful microorganisms developing. For example, instead of maize in the crop phase, a rotation of maize, sorghum and millet may be followed, so that no one host-specific group of microorganisms has a chance to build up year after year.

Even on farms which follow a ley or fallow crop rotation, some chemicals and mechanical means of maintaining fertility are often used. The decision on which system is 'best' is governed by a number of factors such as soil type, cropping history, rainfall distribution, labour and chemical costs. Thus there is no universal 'best' system – the unique situation of each farm will determine the emphasis given to the two broad systems of maintaining fertility. The role of chemical sprays to reduce weeding labour is acquiring increasing importance. Their use involves relatively skilled trained workers. In essence:

a rotation need not necessarily involve the use of a legume or fallow during the 'resting' phase between crops;

substitutes which do virtually the same job as a legume phase usually work in practice, but their profitability relative to legume leys or fallows is still being evaluated in most regions.

Table 11.2 has a summary of how chemical and mechanical methods can benefit soil fertility.

Economic aspects of crops and crop rotations

A crop rotation is a pattern or sequence of plantings of the same or different types of crops along with 'rest and recovery' periods for the land. A rotation can mean anything from a single 2–3 year crop with 10 year bush fallow rotation, a flexible, opportunistic product combination over a very short period (1–2 months); or, any time period with crop, ley or fallow combination in between these extremes. Regardless of length, all rotations have a common purpose, viz: to maintain or increase the level of soil fertility and hence level of production.

A rotation may be designed to maintain or increase the yields of crops by exploiting a 'complementary' relationship with another agronomic activity. For example, a legume crop might be grown in a cycle of non-legume crops in order to restore nitrogen levels in the soil. Or a fallow period might be used in a rotation to conserve moisture or to break crop disease cycles. We are focussing here on such questions as:

how to analyse the economics of different cropping rotation systems;

how to use farm management analyses for planning what to do next time;

what economic techniques to use, and in what circumstances.

First, for any sort of planning it is essential to get a thorough idea of what is involved in the growing of any particular crop. The main pieces of relevant technical and economic information about a crop have to be searched out before examining 'where and how' a crop might fit into the farm plan. For an excellent illustration of what should be known before any planning 'tricks' can be used, see the following activity budget (Table 11.3; taken from J. B. Hardaker's PhD thesis, University of New England, Australia, 1975).

Now, we want to look more closely at one of the key aspects of all cropping decisions – the rotation. The following information is needed when considering crop rotations:

minimum period of the ley or fallow phase to restore fertility;

Table 11.3. *Activity budget for sweet potatoes in Tonga in 1974*

1. Definition
Local name: kumala
Scientific name: *Ipomoea batatas*
Grown as a staple using 'traditional' technology
Local varieties

2. Seasonality
(a) Planting dates: normally planted between March and
    October, but can be planted year round.
(b) Growth period: 4–7 months according to weather
    conditions, etc., typically 5 months.
(c) In-ground storage: harvest can be delayed for up to
    2 months without appreciable yield loss.

3. Rotational considerations
(a) Crop sequences: commonly grown after yams or taro, or
    as a first crop after fallow on less fertile land. Usually
    followed by cassava or fallow. Not recommended to be
    grown on same area in succession.
(b) Intercropping: may be grown as an intercrop in young
    bananas (effective area 33%).
(c) Soil fertility considerations: high levels of soil nitrogen
    may cause excessive vegetative growth and poor tuber
    production.

4. Planting
(a) Spacing: typically planted about 1 m × 1 m (grown as a
    row crop under mechanisation).
(b) Planting material: grown from stem cuttings about 30 cm
    long, three or four per hill. 0.05 ha will provide enough
    planting material for 1 ha.

5. Other inputs
Fertilisers are not used. Dusting against weevil borer is
recommended but seldom practised.

6. Labour requirements

Job	Man hours/ha
Prepare planting materials	60
Plant	100
Form hills	100
Weeding (months after planting): 1	75
2	55
3	35
Harvest	345

7. Yield
Average: 12.5 t/ha

8. Nutritional aspects
Usually consumed boiled or baked. Contains 4.2 MJ/kg
edible portion, 1.5% protein, 15% waste. Not a preferred
staple – maximum of 35% of energy intake from this source.
Short post-harvest storage life.

9. Marketing
Typical local price in 1974 of $5.50/100 kg, net of selling
costs.

maximum period of crop phase;

feed supply for animals from the rotation;

gross margin from crops over successive years,
assuming 'proper' technology;

whether bush or dry fallow is essential to grow an
acceptable crop.

Most important: at first sight, it would seem that the
application of the law of diminishing returns to the
problem of declining yields would suggest that the field
be continually planted to the crop until the crop gross
margin fell to the level of the gross margin obtained
from the restorative fallow or ley phase. But in
deciding the best rotation, the relative lengths and
gross margins of each phase of the rotation have to be
taken into account. The average gross margins for the
whole rotation need to be compared, not just the gross

margins of the individual phases of the rotation. The
examples in the next section are given to reinforce this
point.

Analysis of simple two-phase rotations

In practice, a farm following a rotation of 4 years of
fallow and 6 years of crop would have 4/10 of its area in
fallow and 6/10 in crop, rather than having the whole
area under fallow for 4 years and crop for the next
6 years. Similarly, a farm following a 5 years of fallow,
3 years of crop rotation would have 5/8 of its area
under fallow and 3/8 of its area under crop.

   To calculate the gross margin per typical hectare
after a rotation has been set, it is necessary to multiply
the gross margin per hectare of the fallow phase by the
proportionate area devoted to fallow and add to it the

Table 11.4. *8 year, whole farm rotation*

Year	Field Number							
	1	2	3	4	5	6	7	8
1	F1	F2	F3	F4	F5	C1	C2	C3
2	F2	F3	F4	F5	C1	C2	C3	F1
3	F3	F4	F5	C1	C2	C3	F1	F2
4	F4	F5	C1	C2	C3	F1	F2	F3
5	F5	C1	C2	C3	F1	F2	F3	F4
6	C1	C2	C3	F1	F2	F3	F4	F5
7	C2	C3	F1	F2	F3	F4	F5	C1
8	C3	F1	F2	F3	F4	F5	C1	C2

product of gross margin per hectare from crop and its proportionate area. For example:

GM fallow/ha = $16 (5 years in 8)
GM crop/ha = $64 (3 years in 8)
GM/ha/yr from rotation =

$$\$\frac{5}{8}\times\frac{16}{1} + \frac{3}{8}\times\frac{64}{1} = \$10 + \$24 = \$34$$

Rotations in tabular form

The 5 year fallow or ley, followed by 3 year crop, rotation can be shown in tabular form. Let us say the farm is 8 ha in area. In the 8 year rotation, each area carrying a particular stage of fallow or ley crop would be 1 ha. The same piece of land has fallow for 5 years and crop for 3 years. The 8 year cycle for the farm as a whole is shown in Table 11.4. F1, F2 etc refers to 1st year, 2nd year fallow and so on. C1, C2, C3 refer to 1st, 2nd and 3rd year crop.

At the start of the 2nd year, the following events take place:

Area 8 (which has been cropped for 3 years) becomes fallow (it will remain in fallow for 5 years).

Area 5 (which has carried fallow for 5 years) is ploughed up and sown to crop.

Area 6 is cropped for the second time, and field 7 cropped for the third time in a row.

Areas 1, 2, 3 and 4 remain in fallow, which advances in age by 1 year.

A similar process is repeated at the start of each of the next 6 years. The profitability of a cropping rotation depends on:

the GM of the crop phase;

the GM of the non-crop 'restorative' phase;

the relative lengths of the 2 or more phases of the cropping rotation.

To show the importance of analysing the crop rotation as a whole, rather than taking the gross margins of the individual phases as a guide to action, we will now analyse what might happen if the crop phase is extended, and the fallow phase reduced. Here, the rotation will be 3 years fallow and 5 years crop. If the land is cropped for 5 years in a row, rather than for just 3 years as before, the average gross margin from the cropping phase of 5 years will probably be less than the average gross margin from the 3 years of crop. The reasons for this are that there will be a decline in fertility, higher variable costs for fertiliser, weeding and sprays; also, average yield will be lower. Thus the gross margins for the new rotation might be $16 for the fallow phase, and $40 for the 5 years crop phase (before, the annual average GM for the 3 year crop phase was $64). So, the annual GM/ha/yr for the new rotation will be:

$$\frac{3}{8}\times\frac{16}{1} + \frac{5}{8}\times\frac{40}{1} = \$6 + \$25 = \$31$$

This is $3 less than the GM/ha/yr for the original rotation, even though the GM from crop ($40) is still two and a half times that from the fallow ($16).

Here, *more* crop leads to a lower average GM for the new rotation. This seems to be an anomalous conclusion, although it is correct. It would be easy to think as follows: 'crop has a higher GM than fallow; therefore, the longer the land can be kept in the crop phase, the more profitable the rotation will be'. So, some important conclusions on the economics of crop rotation are that:

lengthening the phase of the enterprise with higher GM does not necessarily lead to higher relative profits;

extreme caution must be exercised when lengthening the cropping phase (it is necessary to ensure that the average GM from the crop over the longer rotation does not fall too much in relation to the average crop GM in the shorter rotation);

considerable effort should be made to ensure that the GM from the fallow phase remains as high as possible (on short rotations, a rise in fallow GM can have a significant effect on income).

So far we have looked at some very simple cases involving only one crop and a fallow or ley.

Rotations in horticultural crops

Horticultural cropping is usually complex. There can be many crops, some complementary, which also may tie up land for several different time periods within a year. The way to handle these situations is to calculate the total gross margin (TGM) per rotation per time period, instead of per hectare per year. Consider the yearly pattern of production where crops *X*, *Y* and *Z* are grown together on 1 ha of land two times during a year, and crops *A* and *B* are grown once in the year.

$X = 0.5$ ha	$A = 0.6$ ha
$Y = 0.3$ ha	$B = 0.4$ ha
$Z = 0.2$ ha	

The rotation is as follows:

Month	Land use (per ha)	Period
Nov.–Dec.	Fallow	1
Jan.–Feb.	X 0.5 Y 0.3 Z 0.2	2
Mar.–Apr.	Fallow	3
May–June	X 0.5 Y 0.3 Z 0.2	4
July–Aug.	Fallow	5
Sept.–Oct.	A 0.6 B 0.4	6
Nov.–Dec.	Fallow	1

So, between November and the end of June there is a fallow, then a cropping sequence of 3 crops together in January and February, a fallow for another couple of months, and then a 3-crop sequence from early May to late June. The land is fallowed for 2 months, and after this it is cropped for 2 months. Then, back to the start of the rotation in early November. The economics of the rotations are analysed as follows:

Rotation I	Period[a]
Fallow (Nov.–Dec.)	1
Crops *X*, *Y*, *Z*	2
Fallow	3
Crops *X*, *Y*, *Z*	4

[a] Total period: 8 months

Rotation II	Period[a]
Fallow	5
Crops *A* and *B*	6

[a] Total period: 4 months.

The land now goes back to fallow at the end of this 12 month cropping cycle. To calculate the gross margin of the 12 month crop rotation, the GMs of rotations I and II are added. Thus:

(*a*) *Rotation I*

	(GM($))		
Crop	X	Y	Z
1st phase	100	50	200
2nd phase	80	160	150
Total	180	210	350

Total GM from rotation I = \$740
GM/month (8) = 740/8 = \$92.5

(b) Rotation II

Crop	(GM($))	
	A	*B*
One phase	300	150

Total GM from rotation II = $450
GM/month (4) = 450/4 = $112.5
TGM for year-long
rotation = 740 + 450 = $1190

By examining the GMs of each rotation profits may be increased by changing:

the mix of crops within each rotation;

the blend of crops between each rotation.

Mixed cropping

Budgeting lets the crop grower and his adviser experiment with different plans on paper. Planning for increasing the revenue of a mixed cropping farm does not always involve a complete reshuffle of all activities. In many cases, a farmer wants to know the likely costs and benefits, on the business as a whole, of a single change. Here, partial budgeting is a valuable aid to planning. At other times, he may be interested in knowing the effect on profits of a change in yield, cost or selling price, in one or more of the activities on his property. In this case, parametric budgets and break-even budgets are of value. Where changes require examination of the effect of variations in more than one of the activities, then gross margins, budgeting, simplified programming, and (for more complex situations) linear programming and systems analysis are appropriate tools. We have already discussed the use of partial and parametric budgets; here we will consider gross margins budgets and simplified programming as bases for selecting a profitable cropping plan.

Gross margins budgets can be useful starting points for decisions on crop farms, as long as their limitations are kept in mind and proposed changes are looked at 'in total' (see Chapter 8). Table 11.5 is an example gross

Table 11.5. *Gross margins for tomatoes*

	$/ha
Returns	
25 tonne per ha @ $60/tonne net, on farm	1500
Variable costs	
Seed = 280 g/ha @ $16/100 g	45
Fertiliser – 0.5 tonne/ha complete 5 : 2 : 1	100
Pest and disease sprays	80
Irrigation	25
Harvest labour @ $10/tonne	250
Cultivation, planting and weeding	480
Total	980
Gross margin per hectare	520

margin calculation for vegetable crop production where some hired labour is used.

The analysis of a mixed-cropping farm is concerned with individual activities and also how they are combined. Activities may be related to each other in a competitive, supplementary or complementary manner (see Chapter 4 for details). Such relationships should be considered in the analysis, e.g. cash crops which are sown and harvested at the same time are competitive for labour, equipment and land. If a large number of complementary or supplementary activities exist, then it is possible to use most of the resources most of the time.

In planning mixed cropping, a good grasp of the connections between crops and inputs, between types of crops, and between the stage-by-stage (production, harvesting, marketing) decisions is needed. The technical needs of a possible new activity must be thought out properly. Also, when considering expanding the area cropped there is a chance that the extra area can stretch resources, especially labour and management, to the point where total production is in fact reduced. Delays, inadequate seed-bed preparation, weeds, moisture stress due to delayed irrigation, or poor crop husbandry at the critical periods, can reduce the return from a larger area to below that from a better managed, smaller area.

The first step when planning with gross margins is to examine each activity to see whether its present gross margin could be improved by using better technology or management. If activity *A* has a gross margin of $100/ha and activity *B* a gross margin of $60/ha, it is wrong to infer that the most profitable step to take would be to expand activity *A* at the expense of activity *B*. A simple change of technique such as a new plant variety or weedicide or time of planting may lift the gross margin of activity *B* to $120, whereas there may be no known way of improving the performance of activity *A*. Present gross margins can be used as a base for deciding on future action when:

the scope for applying improved technology to each activity has been properly assessed;

the physical and financial limitations to the expansion of each activity are known;

capital is available to finance any change.

Where the resources used for different activities are similar, the gross margin per hectare is a simple and useful comparative measure. Gross margins from alternative crops have to be expressed in terms of a common resource so that the profitability of the crops can be compared. Crop gross margin per unit of land area, per hour of labour, or per dollar of capital invested, are commonly used as a basis of comparison. To make most profits, first get the most from that resource which is the most limiting. An example of gross margin planning, using this approach, follows.

Pick the crop which gives the highest gross margin (GM) per unit of limited resource and expand it until the limiting resource – be it land, seasonal labour or (working) capital – is fully used. Then select the activity with the second highest expected GM, and expand it until it, in turn, meets a restraint on further expansion. Repeat the exercise until most of the available resources are being fruitfully used. If a restraint, say seasonal labour, is met in the activity with the highest GM, then the profitability of hiring (i.e. bringing in) extra seasonal labour to help expand this activity should be looked at. If, after hiring labour, it is still the most profitable, then it should be expanded until the GM of the extra area brought into use for this activity is less than the GM from the next most profitable crop (and so on).

One way to use GMs in planning is to draw up a planning table of the physical and financial resources,

can be used. Restraints in production might be physical limits such as land, labour and plant; financial ones (capital and credit) available; or technical limits such as rotation requirements. Institutional factors such as laws, social rules, water rights, quotas, or personal factors (such as dislike of certain activities or fear of risk), can be further restraints. People who draw up budgets usually use either the average, 'most likely' yield for the activites, or the expected value of yield derived from probabilities. To calculate expected value, we will say that over a 10 year period, the following crop yields might be expected to occur (Table 11.6*a*).

Table 11.6. *Probabilities of crop yields*
(*a*)

Yield (tonnes/ha)	Years in 10	Probability
0.10	1	0.1
0.30	2	0.2
0.50	3	0.3
0.70	2	0.2
0.90	2	0.2

The total expected value is found by multiplying the yield by the probability that such a yield will occur, and summing the results.
(*b*)

Yield/ha	Probability	Expected yield
0.10	0.1	0.01
0.30	0.2	0.06
0.50	0.3	0.15
0.70	0.2	0.14
0.90	0.2	0.18
Expected yield		0.54 tonnes/ha

and to define the highest level (restraint) at which they
Example Suppose that a farmer can follow three possible crop rotations, viz: *A*, *B* and *C*. He considers that land, weeding labour and harvesting labour are the most limiting resources. He has 10 ha of land (all of the same quality), a labour supply of 100 h for weeding, and 90 h available for harvesting. Also, he has decided

Table 11.7. *Resource table*

(1)[a] Resources	Unit	(2)[b] Resource limit	(3)[c] Rotation resource requirements		
			A	B	C
Land	ha	10	1	1	1
Weeding labour	hours	100	25	16	0
Harvest labour	hours	90	10	5	10
Rotation (3 ha in 10 fallow)		1	1	1	
Expected gross margin ($/ha/rotation)		200	190	130	

[a] Column 1 contains the resources that are available to be used.

[b] Column 2 has a list of the maximum amount of each resource which is available.

[c] Column 3 contains the amount of each resource which each crop rotation uses, e.g. 1 ha of rotation A uses 25 h of weeding labour, 10 h of harvest labour and 1 ha of the 7 ha available for cropping once the rotation restraint of 7 : 3 is considered. It also shows the gross margin of each rotation.

would need. His adviser has helped him to work out the expected gross margins from each crop.

With this information, his adviser could use the technique known as simplified programming to help him to work out the crop plan which is most likely to produce most profit. (The same could be done using food/ha instead of gross margin/ha in dollars.) how he could follow each of the rotations A, B, and C, and the amounts of weeding and harvesting labour he

A planning sheet is drawn up (Table 11.8) and rotations selected on the basis of the highest gross margin per unit of the assumed most limited resource, say, land. Crop rotation A is expanded to the limit permitted by the planning constraints. Then other activities are progressively introduced until as many resources as possible are fully used up. In Table 11.8 we show how this is done. Crop rotation A has been expanded to the limit (4 ha) where all weeding labour is used up. But, there are still 3 ha of land which could be

cropped (remember there was only 7 ha available to crop; 3 ha was in fallow). There are also 50 h of harvest labour remaining to be used. The farmer cannot follow crop rotation B because there is no weeding labour available, but he can follow crop rotation C, on the remaining 3 ha of land. The farm plan is 4 ha of crop rotation A, 3 ha of crop rotation C, and 3 ha of fallow. Total gross margin is $1190. There are 35 h of harvesting labour unused. As weeding labour appears to be quite a limiting resource, it may be worthwhile to redo the farm plan, selecting crops on the basis of gross margin per hour of weeding labour required. We will have a look at this plan in Table 11.9.

After following the same steps, the second plan involves growing 6.25 ha of crop rotation B, 0.75 ha of crop rotation C, plus 3 ha of fallow, for a total gross margin of $1285. This is slightly better than the first plan, also there are 5 h of harvesting labour unused.

With both plans, the shortage of weeding labour stops the growing of more of either crop rotation A, or crop rotation B, both of which are more profitable than crop rotation C. However, these plans are not necessarily 'the answer' as to what the farmer should do, but rather information which he can use when deciding what to do. In practice he would investigate the possibilities of selling his surplus harvest labour and buying in some weeding labour to enable more of crop rotation A to be grown, or for crop rotation B to be expanded to the full 7 ha.

There are other things to think about too. For instance, there may not be much difference between the expected income from the two alternative plans. Immediate issues to consider then are: which crop rotation does the farmer feel is the more likely to be a success, which is the less risky, which crop rotation does he grow best, which meets the greatest diversity of his needs, how best to use the surplus harvesting labour?

In summary, the two plans selected on the basis of GM/ha, and the other on GM/h of weeding labour are (in this case) of similar profitability. In practice, because there is little difference in this case between the total gross margin generated by the two plans, either plan could be 'best', depending upon how highly the farmer rates the other factors which come into the decision.

Table 11.8. *Planning sheet one*

Resource	Available	Select rotation A (Highest GM/ha)			Select rotation B			Select rotation C			Final Balance	
		Maximum level	Used	Balance	Maximum level	Used	Balance	Maximum level	Used	Balance	Total used	Balance
Total land (ha)	10	10/1 = 10 ha	4 ha	6 ha	6/1 = 6 ha	0	6 ha	6/1 = 6 ha	3 ha	3 ha	7 ha	3 ha
Weeding labour (h)	100	100/25 = **4 ha**[a]	(4 × 25) 100	0	0	0	0	0	0	0	100	0
Harvest labour (h)	90	90/10 = 9 ha	(4 × 10) 40	50 h	50/5 = 10 ha	0	50 h	50/10 = 5 ha	(3 × 5) 15	35	55 h	35 h
Rotation constraint (ha)	3/10 ha fallow, 7/10 cropped	7 ha	4 ha	3 ha	3/1 = 3 ha	0	3 ha	3/1 = **3 ha**[a]	3 ha	0	3 ha	0
Gross margin/ha		$200						$130				
TGM/Rotation		$200 × 4 = $800						$130 × 3 = $390				
TGM all crops	$1190											

[a] Figures in bold represent most limiting factors, used to maximum.

The whole farm approach has produced many planning techniques, the objectives of which are to find the 'best' plan. It is important to keep in perspective just what really determines the level of success of a farm operation. Experience shows that plans may not be as important as how any reasonably sound plan is put into action. With linear programming, which is now a major computer-based development of the simplified planning technique described here, the profitability of quite a number of different plans is often shown to be not widely different from the 'best'. What is important in this circumstance is the level of technical efficiency in putting any of these plans into action. Often, formal planning in the textbook sense is less important than:

achieving a good performance within the chosen plan, e.g. crop yield and stock performance;

intensity of production, e.g. level of high return enterprises;

keeping high-cost inputs at a low level (particularly machinery, chemicals and paid labour) within the chosen plan.

Nevertheless, for the intensified and diversified farm, techniques considering the return to all resources, rather than to a particular resource, are appropriate planning tools. In practice, most farmers do not think in terms of 'farm planning by linear programming', but are applying the same economic principles (without fine tuning) when they use partial budgets and gross margins to select and test farm plans. Analysis and planning of mixed farming enterprises requires that the technical relationships between inputs and outputs, and between different outputs, be well known. Mixed-farm planning techniques are based on gross margins, limits to production and the principle of substitution.

Variability of returns

Diversification can be a way of reducing the degree to which income can fluctuate. Crops with different growing and marketing seasons can be combined in such a way that possible low returns from one crop can be compensated by high returns from another crop. An example of the different effects of various combinations of crops with gross margins variations is shown in Table 11.10

If the crop grower has planted the whole 1 ha to tomatoes, which has the higher gross margin, total gross margin would have fluctuated between $1500 and $530 and averaged $1025.

If half of the land was planted to potatoes the total gross margin (TGM) would have fluctuated from $900 to $655 and averaged $787 – variation is less than would have occurred with just one crop, so too is average TGM. This is the 'cost' of stabilising the income from the 1 ha of vegetable land.

Flexible production and marketing plans, which can be altered at short notice, can help to reduce the extent by which returns may at times plummet; they also make it possible to cash in on opportunities as they arise. The flexibility which an enterprise has depends a lot on the types of crop grown. Changes in the vegetables which are grown can be made much more readily than changes to tree crops, for example.

The perishability and storability of a crop also affects the flexibility a farmer has to change his production. Storing crops allows greater flexibility in harvesting and marketing decisions. As well, with crops having alternative end uses, some products can be used for the fresh or processed market, some may have an end use even if damaged.

Summary of planning mixed farms

The more complex budgeting techniques (such as linear programming) include many variables and production constraints in forming a plan. Other methods take account of the variability of possible prices, yields, and climatic conditions. These techniques can come up with economic 'best' crop combinations. They must then be evaluated in the light of the real-life factors which were too difficult to include in the model.

Gross margins alone will not indicate the best enterprise combinations on an individual crop holding; GMs are an easy-to-use guide to choosing crop activities, as long as other factors such as product-price stability, yield variability and technical and management requirements are given careful consideration.

Table 11.9. *Planning sheet two*

Resource	Available	Select rotation B			Select rotation C			Select rotation A			Final balance	
		Maximum level	Used	Balance	Maximum level	Used	Balance	Maximum level	Used	Balance	Total used	Balance
Total land (ha)	10	10/1 = 10 ha	6.25	3.75	3.75 ha	0.75	3		Surplus		7	3
Weeding labour (h)	100	100/16 = **6.25 ha**[a]	6.25	0	0	0	0		Limit		100h	0
Harvest labour (h)	90	90/5 = 18 ha	6.25 × 5 = 31.25	59	59/5 = 14.8 ha	0.75 × 10 = 7.5	51		Surplus		39h	51h
Rotation constraint (ha)	3/10 fallow, 7/10 cropped	7 ha	6.25	0.75	**0.75 ha**[a]	0.75	0	0	Limit		7ha	0
GM/ha		$190			$130			$200				
GM/hr weeding labour		$11.90			—			$8.00				
TGM/rotation		$190 × 6 = $1187.50			0.75 × $130 = $97.50			—				
TGM all crops	$1285 (1187.50 + 97.50)											

[a] Figures in bold represent most limiting factors, used to maximum.

Table 11.10. *Crop combinations and gross margins*

	Year				
	1 (GM/ ha)	2 (GM/ ha)	3 (GM/ ha)	4 (GM/ ha)	Average
Crop					
Tomatoes	1000	530	1500	1100	1025
Potatoes	500	800	300	600	550
Total Gross Margin					
1 ha Tomatoes	1000	530	1500	1100	1025
0.5 ha Tomatoes	500	265	750	550	512
0.5 ha Potatoes	250	400	150	300	275
Combined total	750	655	900	850	787

Harvesting and marketing

Crop growers, especially vegetable producers, face some major uncertainties which impose difficult planning problems. Produce price varies often from day to day, and from the corresponding periods of one year to the next; yields vary from season to season; planting and harvesting times may vary considerably due to bad weather – this affects planning of other crops; how much labour is wanted, and when, depends on the crop husbandry needed and the size of the harvest.

A. N. Rae, in his study of the management-decision problems of vegetable producers looked at the stage-by-stage decision process in vegetable production – he calls it a 'sequential-decision process'. It is shown in Fig. 11.1. He considers that one important management task of a vegetable farmer is to:

. . . attempt to 'interlock' the decision flows for all crops produced concurrently so as to minimise the occurrence of resource bottlenecks. All stages of the production process require the input of cash and other resources – only after the crop has been marketed is a cash return received.

. . . production of a crop may be terminated at some point before distribution, as indicated in the figure, depending on the decision-maker's expectations of future returns from marketing the crop, compared with the likely future outlay of cash required to bring the crop to the marketing stage, and his expectations of net pay-offs from subsequent crops.

Decisions on harvesting and marketing of crops

One of the purposes for analysing crop activities, apart from obtaining the gross margin, is to distinguish

Table 11.11. *Per hectare costs*

	Crop A ($)	Crop B ($)
Growing	60	120
Harvesting and marketing	120	60
Total variable costs	180	180

between pre-harvest costs and 'harvesting and marketing' costs. Different, but alternative, crops which may have the same total variable costs, can in fact have different dollar losses if both fail to reach maturity. How the total variable costs are distributed between growing and harvesting costs is an important factor in deciding which crop to grow. When a farmer knows the size of these two sets of costs, he is in a position to make better decisions about future cropping activities.

As an example, assume that there are two crops, one of which (crop *A*) has high harvesting and marketing costs (labour, storage, processing, transport), but low growing costs. Crop *B* has relatively low harvesting and marketing costs, but high growing costs. With average yields, the total variable costs (growing, harvesting and marketing) are the same. Table 11.11 contains the relative costs per hectare.

If the farmer took account only of the total variable costs in his planning, he would be likely to draw the wrong conclusions. Since both crops have a total variable cost of $180, he might conclude that there would be no difference in the amount of money which would be lost if the crop failed to grow to maturity. But in many cases, crops reach a stage close to harvesting, and then cannot be harvested because of unfavourable weather, damage caused by insects and disease, or because there is no market. In such cases, the farmer loses only $60/ha with crop *A*, but $120/ha with crop *B*.

Assume that the two crops have the same gross margin and total variable costs in a 'normal' year, but differing growing and harvesting and marketing costs. The figures given in Table 11.12 (p. 111) show how the GMs will change if the yields differ from the normal. Probabilities of departures from the normal can be assigned to these GMs, and the expected values calculated. Thus, crop *B* is better in a good year, equal

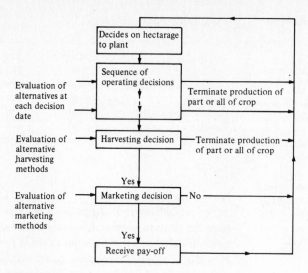

Fig. 11.1. The sequential decision-making process in vegetable production. Source: A. N. Rae, PhD thesis, University of New England, Armidale, NSW, Australia (1970).

in a normal year, and worse in years where there are low or no yields.

Deciding whether to harvest a crop

Once the stage of harvesting a crop has been reached, the main economic factors which determine whether or not it should be harvested for market (not home consumption) are the expected harvesting and marketing costs, the gross income, and the costs of, or income from, the crop residue if the crop is not harvested.

The fact that money has already been spent on growing costs is, at this stage, not relevant to the decision of whether or not to harvest. (This is true even when the total variable costs are more than the total gross income – such a situation calls for cutting your losses rather than making most profits).

If there are no extra costs incurred when the crop is not harvested (e.g. having to chop it up and work it into the soil), or if no income can be earned from the unharvested crop from, say, feeding it to animals, then the decision rule is that the crop should be harvested if the gross income from the sale of the crop exceeds the costs of harvesting and marketing.

When there is either a cost or an income associated with the unharvested crop, this decision rule has to be modified. For example, a farmer may face the following situation, due to a fall in prices, of a perishable vegetable crop which he has grown:

Pre-harvest growing costs =	$100/ha (A)
Harvest and market costs =	$300/ha (B)
Expected income =	$250/ha (C)
Extra cost of disposing of crop residue if crop not harvested =	$100/ha (D)

If he does not harvest the crop, his net costs are $200, i.e. (A) + (D). If he harvests the crop, his net costs are $150, i.e. (A) + (B) − (C). So he is better off by $50 to harvest in this case, even though the harvest and marketing costs exceed the expected crop income. As the $100 growing cost is common to both calculations, it can be ignored (hence the statement made above that growing costs are irrelevant to the decision, at harvest time, of whether or not to harvest). The decision rule which can be deduced from this example is that if the harvest and marketing costs minus the income from the sale of the crop is less than the cost of disposing of the unharvested crop, then the crop should be harvested, regardless of the amount of money spent on growing costs.

The next example deals with the case where there is

Table 11.12. *Possible production outcomes*[a]

Season	Crop A (GM) Low growing costs, high harvest and market costs	Crop B (GM) High growing costs, low harvest and market costs
Normal yield	300	300
50% better than normal yield	480	520
50% worse than normal yield	120	90
Crop failure	− 60	− 120

[a] The GM figures in this table have been calculated from the data used to compile Table 11.10.

income, rather than costs, associated with not harvesting the crop, e.g. where the unharvested crop can be used for fattening animals.

Harvest and market costs = $300/ha ($A$)

Expected income = $350/ha ($B$)

Income if not harvested but fed to animals = $150/ha ($C$)

Profit from harvesting = $50/ha ($B-A$)

Profit from not harvesting = $150/ha

Here, the rule is, if the income from selling the crop minus the harvesting and marketing cost is less than the income from using the unharvested crop residue, then the crop should not be harvested. We can express the decision rule as a formula, where:

GI = Gross income
H = Harvest and marketing cost
D = Disposal cost of (or income from) crop if harvested

Thus if $GI - H$ is less than D, then do not harvest; if it is more than D, then do harvest. Substituting from the second example,

$350 − $300 = $50 < $150, do not harvest.

Marketing*

Vegetable growers usually have a choice of different ways of making use of their product. If not consumed

*See also J. C. Abbott & J. P. Makeham, *Agricultural Economics and Marketing in the Tropics*, Longman, 1979.

by the farm family, produce might be purchased by marketing boards or cooperatives, it might be sold fresh to wholesalers, retailers, or even direct to customers, or it might be sold for processing. Growers can be in the position of making packaging, grading, storage and pricing decisions. Often they can choose whether or not to form or join growers' cooperatives. There is a strong incentive for them to know how their markets work, and to have some fairly positive ideas about near-future and longer-term price levels.

Prices of fresh vegetables fluctuate a great deal due to their perishability, seasonal production patterns, variability in areas grown and in yields, combined with a fairly fixed demand. Fresh vegetables are sold mainly through local markets. Production for processing is often based on contracted prices for certain quantities and qualities. These prices are more stable but generally lower than fresh market prices.

A flexible marketing strategy, using some contractual arrangements, some out-of-season production where there is a relative production advantage, and a number of fresh produce outlets, can help to reduce the harm caused by wide price variations. Meeting a specialist product or quality market can stabilise prices received, but to do so involves higher levels of management, processing and equipment.

Tree crops

The term 'tree crop' is used here to include orchards, woodlots and plantations. Tree crops have a number of common biological and economic features.

Biological features

Each crop has a limited life and so has to be replanted. There is usually a number of years' time lag between planting and the first harvest. There is a succession of harvests from the time of first harvest until the end of the useful life of the crop. Such harvests need not be annual, e.g. thinnings from pine plantations, but in the case of fruit, nut and palm trees, they usually are. Annual operations have to be performed to protect or to promote the growth of the crop (e.g. insect control, fertilising), and to harvest the produce from the crops (e.g. picking the fruit, collecting and transporting). Quite often annual crops and/or pastures are grown between the tree crop, especially in the early years.

Economic features

Tree crops have costs associated with them which are classified as capital, overhead and variable costs. In the time between first spending the capital to establish the tree crop and the first harvest, the farmer does not know for certain what the costs, yield and prices will be. The longer the period to first pay-off, the greater the uncertainty and potential variation associated with prices and costs. To illustrate the main costs associated with tree crops, consider a plantation activity. First, some capital costs of establishment:

Land preparation	Irrigation (plant and lay-out)
Tree purchase	Buildings
Planting	Machinery and plant
Tracks	Weeding and pest control
Fencing	Maintenance during the early years.

Second, the annual overhead or 'fixed' costs which have little or no effect upon the level of yields.

administration, depreciation, insurance;

permanent labour, interest

repairs and maintenance to fixed structure, fire control.

Third, the variable costs, which are directly related to yields. In the early years of a crop's life there may be no production, but the variable costs incurred in these years relate to the output once the plants become productive. In other words, the pre-harvest costs of irrigation, sprays and pruning in the early years, are responsible for the level of yield once the trees begin production. The main pre-harvest variable costs are:

Pruning	Green cover crop (if used)
Thinning	Wiring and propping (if necessary)
Fertiliser	Cultivation
Sprays	Vehicle and machinery costs
	Casual wages

In the following economic analyses of plantation crops we look at a cash flow pattern from establishing a tree crop, and how to decide on tree replacement policy.

Example – tree crop establishment and replacement The following cash flow budget, Table 11.13, with no charges made for interest and tax, has a typical cash flow pattern for establishing tree crops. The costs and returns are in terms of today's dollars. The example deals with establishing 1 ha of 500 trees. The assumptions are:

land cleared and fenced;

land already owned;

a source of irrigation water exists;

to set up an irrigation system using micro jets costs $2.50/tree.

As a tree crop ages, the annual yield eventually declines. There is a yield at which it is more profitable to replant and wait for the new crop rather than suffer the consequences of declining yield from the existing crop. Cumulative cash flows from a perennial crop usually follow the pattern shown in Figure 11.2. The important decision to consider is the year at which the stand should be replaced with a new one, if the tree production is following the pattern outlined. The best time to replace the perennial crop plants depends upon their annual contribution to the cumulative net cash flow from the plantation. We compare the profitability of another year of the present rotation with the average

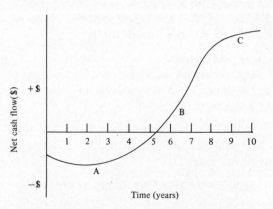

Fig. 11.2. Cumulative net cash flow for a tree crop. The three critical stages are: A, cumulative net cash flow negative at first (establishment costs exceed cash receipts); B, yields increase over time and cumulative cash balance becomes positive at an increasing rate; C, yield stabilises and then falls (cumulative net cash flow continues to increase but at a decreasing rate).

Table 11.13. *Per hectare net cash flow for fruit orchard ($)*[a]

Year	1	2	3	4	5	6	7	8	9
(a) Expenditure									
Capital costs (500 trees)									
(i) Buy tree ($3)	−3250								
(ii) Plant tree ($1)									
(iii) Install irrigation ($2.50)									
Variable costs									
(i) Growing cost ($2.50)	−500	−500	−500	−1400	−1400	−1400	−1400	−1400	−1400
(ii) Harvest & marketing ($2.50 for each 30 kg)				625	−937	−1250	−2200	−3750	−3750
(b) Income									
Production (kilograms/tree)	0	0	0	15	22.5	30	60	90	90
$/ha ($6 for each 30 kg)				+1500	+2250	+3000	+6000	+9000	+9000
Annual net cash flow ($B-A$)	−3750	−500	−500	−525	−87	+350	+2400	+3850	+3850
Cumulative net cash flow	−3750	−4250	−4750	−5275	−5362	−5012	−2612	+1238	+5088

[a] It takes 8 years before cumulative cash flow becomes positive, which is also the first year of full operation. This cash flow pattern is typical of many plantation cropping enterprises. No interest has been charged. Had it been, it would take 9–10 years before the project showed a positive bank balance.

yearly profitability expected to be generated from the full rotation of a new plantation.

The decision rule is to keep the plants in production for another year if next year's expected net cash flow is greater than the annual average net cash flow which could be earned from the next rotation. If this is not the case, the producer will be better off starting a new rotation and reaping the higher average return. Also the new rotation will probably incorporate new techniques, new varieties, new designs which might facilitate cost cutting, or achieve yield or price improvements.

A practical point is that an entire plantation probably will not be replaced all at once. More likely, individual blocks of trees will be replaced at various times in order to spread capital expenditure, and to have a more even pattern of income over time.

This simple rule for perennial crop replacement would be satisfactory if a dollar of costs or revenue is no different in value to a dollar of cost or revenue in the future.

In reality a dollar in the future will have a different value to today's dollar* because of alternative oppor-

*In this section we use concepts explained in Chapter 9, pp. 67–77, under the heading 'Time is Money'.

tunity investments and earning rates. Different 'investors' will have differing preferences for a future versus a present dollar.

So it is necessary to discount net cash flows when appraising projects with a life longer than three to four years. It is necessary to have a rule for replacement which takes account of the value of distant cash flows. The rule is that the tree crop should be replaced when the extra or marginal net revenue from the present rotation is just slightly below the highest amortised present value of expected revenues from the next rotation.

The term 'amortised present value' needs explaining. We showed in Chapter 9 that the present value of a project was obtained by:

calculating the net cash flows which occur in each year;

discounting each annual cash flow to its present value;

adding all the annual discounted cash flows to give a lump sum present value.

The annual cash flows or net revenues received from a tree crop are not even – they are 'minus' or negative in the early years and 'plus' or positive in later years. Amortisation is a technique which converts an uneven stream of net revenue to an even stream which has the same lump sum present value.

Thus, a lump sum present value of $6000 of an *uneven stream* of net revenues per hectare received over 20 years, with a discount rate of 5% is equivalent to receiving an even stream of $480 net revenue per year for 20 years. So the present value of $480 received each year for 20 years is $6000, even though the cash receipts over the period are 20 × $480 = $9600. Another name for an even stream of net cash flows is an 'annuity'. The formula which gives the size of an annuity equivalent to a lump sum present value is known as the 'amortisation formula' (see Appendix 1, Table C).

Thus the replacement rule which takes account of the value of money over time (using discounting), is little different to the earlier one which did not. In both cases we are comparing the return from leaving the stand for one more year with an 'average' or 'representative' annual net cash flow which could be obtained over the life of a replanted stand. The choice of the optimum replanting time depends on assumptions made about yields, costs and prices perhaps 20 or 30 years ahead, as well as on the rate of interest chosen for discounting.

For a better guide to cash flows and optimum replacement time, use expected money values instead of simple averages for the calculation (see Chapter 10 for an explanation of the use of expected money values).

Intercropping

With most tree crops there is a possibility of using the land between the rows, either for cropping or grazing. This is almost always feasible during the early years of any tree crop plantation, when the plants themselves are still small. *However*, a crop between the rows should not extend into the root zone of the trees and rob them of nutrients, and the plants need to be protected from damage by animals. Later, as the plants grow and a canopy forms overhead, the scope for growing crops between the rows is reduced, though grazing of plants that will grow between the rows is usually still possible (as long as the trees are tall).

Growing annual crops between the rows of tree crop is a way of bringing in returns from the land in the early years of establishment of the trees, and, provided that the development of the trees is not impeded, should result in a greater overall return.

Questions

1 What are the main variables which affect crop yields in your area?

2 What practical measures are available to control these variables?

3 What scope is there for legume leys in your area?

4 What is the minimum period under legume ley needed to restore fertility and control weeds?

5 What are the most commonly used rotations in the local agriculture? What *R* values apply?

6 Before any economic analysis can be done you need first to draw up technically detailed activity budgets. Have you all the technical information which is required? Draw up such a budget along the lines of Table 11.3, p. 100. Now calculate the gross margins of (i) the activity and (ii) the rotation.

7 Gross margins, simplified programming and linear programming all apply economic principles to get the most return from the limited resources available. Explain how this is done.

8 How would you decide whether or not to harvest a crop?

9 What steps would you go through in forming expectations about the prices your farmer friends may expect to receive for their production?

10 What is the decision rule for replacing a plantation crop?

11 Use the techniques explained in this chapter to work out the feasibility of some of the local farmers establishing a plantation crop on part of their land which is presently devoted to annual crops.

12 Many people claim that improved crop varieties are 'the answer'. We know of many cases where new varieties which were very promising in trials have been a disaster for the farmers who adopted them. How can such disasters be prevented in the future?

12

Animals

Introduction

Why do farmers in the tropics keep animals? Because animals supply one or more of the following products or services: meat, milk, eggs, fibre (hair and wool), blood and urine, skins and leather, traction and transport, dung, prestige, assets and security in a negotiable form, social and cultural services (e.g. bride price payment); and/or because farmers like them. H. R. Jahnke, in his text *Livestock Production Systems and Livestock Development in Tropical Africa*, published by Kieler Wissenschaftsverlag Vauk in 1982, says that animals perform four functions:

(i) output;

(ii) input;

(iii) asset and security;

(iv) social and cultural.

As output, animals provide food and non-food products for subsistence consumption, cash income from sales, and a means of helping the family to get a balanced, more interesting, diet. The inputs animals provide include farm work and transport, manure, and they can have an integrating function, e.g. using non-arable land within the farm, using otherwise (seasonally) unemployed labour, and converting low-value crop residues to high-value animal products.

Animals have an asset and security function by providing a tradable asset, whereas land often cannot be sold because it belongs to the group or clan. Banks are often neither available nearby nor trusted, but livestock is regarded as a relatively safe and durable means of storing private wealth, which earns 'interest' because it produces offspring each year. The social and cultural function goes past the straight economics of production. For example, the bride price may be payable only in cattle; the camel can be valued for racing ability, not its milk or hair production; a person's, or family's social status can be related to numbers, species and quality of stock owned.

Animal-feeding systems in the tropics range from nomadic grazing of rangelands, through ranching systems where the animals are not moved from their home farm, to integrated crop–livestock farming, and finally to 'factory' farms.

The methods of husbandry of animals in the tropics also vary greatly. At the simplest level of animal production, there is the semi-subsistence household with small numbers of a few species (kept mainly for domestic, not commercial, use). For example, there may be some chickens or ducks, two or three sheep or goats, a pig or two, a growing crocodile, a cow or camel for milk, a horse or donkey for transport, a buffalo or ox for field work and usually a dog for company and protection. Most of the feed for the smaller animals comes from household wastes and grazing close to the compound. The larger animals are usually fed on crop residues. It would be foolish to suggest that exponents of the farm management economics discipline have much, if anything, to say to farm families who run this type of small, mixed, animal operation.

The next unit of production is the medium-sized farm which sells a good deal of its production in the market. Then, there is the 'factory' farm – which typically is relatively large, fully commercial and intensively produces eggs, chickens, pigs and cattle. Some of the techniques of farm management economics can be used in advising the operators of both medium-sized and factory farms.

Animal management economics

The different species in use in the tropics – cattle, buffaloes, elephants, sheep, goats, camels, equines (donkeys, horses and mules), pigs, chickens, ducks, crocodiles and dogs – have different nutritional and management needs. We will concentrate on some of the biological and applied-economic principles which underlie the management of most of them. The three main components of any animal production system are:

(i) the animal, which converts feed into useful ends (be they edible, wearable, or saleable products, or animal power for farm work or general transport);

(ii) feed;

(iii) costs and returns associated with the system.

The animal as a conversion mechanism

The animal is the 'mechanism' by which feed is converted into useful ends; it is the economics of this conversion which is our concern. Two important aspects of the economics of animal production are:

(i) the efficiency with which feed is converted by the animal; and

(ii) the management or production system which is used for the different livestock types and classes.

We will now outline the important aspects of the 'conversion mechanism'.

Health

For our purposes, it should be noted that it is possible to control the major animal disease problems by using vaccines, veterinary services and by paying attention to hygiene and husbandry. Such expenditure is an essential precaution, as the penalty costs of failure to take reasonable care of the stock are usually quite high. In contrast, the relative cost of an effective disease control policy is low.

Most animals carry disease microorganisms and parasites to some extent, but usually not enough for their effects on growth and production to be obvious. Such 'hidden' levels of disease can cause losses in production, and hence have economic costs. The risks of such losses are much greater when feed, water and shelter are not adequate. Bad husbandry can give disease a chance; with good husbandry disease is often prevented, or at least prevented from significantly reducing productivity.

Genotype

It is clear that some members of the one species are genetically superior to others in their ability to convert feed to animal product and in ability to reproduce. Demonstrations of this are seen in the application of modern population genetics to production in the poultry and pig industries, where it has been possible to produce animals with much more efficient food-conversion ratios and reproduction rates than was possible 20 years ago.

Stage of growth

The stage of growth has a big influence on the economic outcome of a feeding operation. In many cases it has a far more economic importance than genotype. The feed-conversion ratio (FCR) is defined as:

$$\frac{\text{kg feed}}{\text{kg livestock gain}}$$

The higher the FCR, the less efficient is the feed conversion process. Older animals have a higher FCR than younger ones.

Suitability to its environment

All animal species have their environmental limitations. Thus it is difficult to farm most sheep in the wet tropics. Also, within the genus, certain species are better adapted to particular environments, e.g. *Bos indicus* (Brahmin type) cattle are better suited to the wet tropics than *Bos taurus* (British breeds). In the subsequent discussion on economic analysis we assume that the animals are suited to the environment.

Exercise

The energy requirements above maintenance increase when the animal expends energy grazing. In feed lots, the exercise factor is approximately 15% above maintenance; when grazing medium-to-good quality pastures it increases to 35%. On poor, sparse pastures the requirement of energy for exercise can rise to as high as 60% above maintenance.

Disease, psychic and environmental stress

Many disease conditions will increase energy requirements, as will the psychological stress which results from excessive crowding of stock in a small area, rough handling and frequent disturbance by humans, dogs or pests. As intensification increases, greater attention needs to be given to psychic aspects of animal performance. Excessively high and low temperatures, or wet and windy conditions, also raise energy needs.

Nutrition

Poor nutrition slows growth, delays the animal reaching reproductive maturity, increases losses of both very young and old stock, reduces reproductive performance (oestrus cycling, semen quality, birth percentage, birth weight, survival rate, reproductive life, reproductive interval) and increases susceptibility to diseases and parasites.

The production system

The type of animals, the different classes of each type, and the way in which they are managed, can be termed the production system. Animal-production systems can be classed as extensive or intensive. In extensive systems of animal production, animals range widely and harvest their own food, and total production from the entire activity is the key to success, e.g. a ranching or herding enterprise. Intensive forms of animal production may be poultry, pig, or some dairying enterprises where the farmer brings the feed to the animal, where disease control is a major factor and where the essential feature is a high level of production per animal.

Within these broad categories there are three major ways in which the flock or herd replaces itself. These 'replacement systems' are:

non-breeding animals, usually replaced with bought-in replacements;

breeding with bought-in female replacements;

self-replacing.

Non-breeding systems include such activities as production of broiler chickens, fattening young cattle, goats, sheep and pigs. Breeding with bought-in female replacements includes buying young laying hens, mated heifer cattle, young female pigs (gilts) or goats. Self-replacing systems of production are slightly more complex than the above two. Examples can be seen in pure breeding using animals of the same breed, or a cross-breeding system to produce first cross breeders. Two important economic questions to be answered whenever one looks at any animal production system are:

how does the system replace itself?

what are the costs, other than feed and husbandry, associated with replacing the system?

Flow charts

To carry out economic analyses of various replacement systems it is useful to make a flow chart for each enterprise which shows which animals, (and in what numbers) go where within the system. That is, to use rates of birth, culling, death and 'selling for old age' (cast for age, cfa) to calculate the numbers of each class of animal within each activity which will be sold and retained. Three examples follow. Note: the techniques can be used in any animal enterprise.

Non-breeding with bought-in replacements

This is the simplest of all the animal-production systems, with young animals being purchased and grown out for sale, or used for some form of production and then sold. In most of these systems there is a mixed age structure, so as to ensure that income flows are more even over time.

Example: Egg-laying hens

Assuming – only 1 year of laying
– flock of 100 birds
– 5% deaths

$$\begin{array}{ccc}
\text{Purchase} & \text{Lay for 1} & \text{Sell 95 cast} \\
\text{100 pullets} \rightarrow & \text{year} \rightarrow & \text{for age hens} \\
& \downarrow & \\
& \text{Sell eggs} &
\end{array}$$

The situation regarding broiler chickens is even more simple than that of egg-laying chickens. These are simply purchased as chicks, fed to grow to an adequate size for broiling, and sold. The same system applies to the fattening of cattle, sheep and pigs.

Example: Fattening of sheep/lambs/pigs

Assuming – 2% deaths

Purchase 50 lambs → Feed to fatten → Sell 49 fattened animals

Bought-in female replacements

The system of using bought-in female replacements is similar to a self-replacing system in that females are mated to breed offspring or to produce eggs. However, instead of keeping a percentage of the female offspring to replace the cast for age (cfa) animals, animals which are ready to breed are purchased when needed. This means that, for any 1 year, all offspring plus the cfa females are sold and a number of young females which are ready to breed are purchased.

An advantage over a self-replacing system is that extra land and feed is available to the herd or flock, as there are no animals being carried to be grown for later breeding. Whilst there is the saving, there is also the cost of purchasing replacements.

Example

– herd of 20 cows;
– cows breed at 3 years old (3 YO), 4 YO, 5 YO and 6 YO;
– assume no deaths, for simplicity;
– 50% of cows calve successfully each year.

5 replacements purchased → 20 cows ↓ 10 calves sold when older → 5 cfa cows (the oldest cows)

Self-replacing systems

Self-replacing systems are the most complex of the three. This system is common in cattle, goat, sheep, pig and dairy activities. In a self-replacing system some of the female offspring are retained to replace cfa animals in the herd or flock. Thus for any herd or flock the age structure ranges from animals which are younger than mating age, to those old enough to be cast for age. Therefore in a 20 cow herd, breeding cows may be aged say 3 years, 4 years, 5 years and 6 years and there will be the heifers nearly ready to mate, and young growing stock. The way in which animals flow through the system is shown in the flow chart.

Example: Cattle (20 cow breeding herd)

Assuming 50% calving, no deaths (in practice, 3–4% deaths occur)

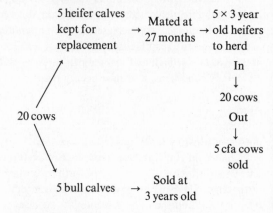

Herd Structure

20 cows (3–6 years)	1 Bull
5 Heifers (2 years)	5 Steers (2 years)
5 Heifers (1 year)	5 Steers (1 year)
5 Heifer calves	5 Bull calves

Example: Goats: (20 doe breeding herd)

Assuming – 100% kidding;
– first kid at 2 years
– sale of kids at 6–7 months
– culls at 5 years after kidding

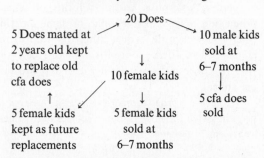

The three animal-production systems (bought-in non-breeding, bought-in female replacements, and self-replacing) have some costs which are specific to each system. Apart from feed and husbandry costs, which are common to all systems, a bought-in non-breeding production system can have as costs (i) deaths and (ii) depreciation due to the purchase price of young animals being more than the sale price of old animals. However, often the value of the animals appreciates.

Table 12.1. *Main cost items in animal systems*

Beef	Goats, sheep	Dairy	Poultry and pigs
Feed			
Fertiliser for pastures or fodder crops			Bought feed
Hay and silage	As for beef	As for beef	Home grown feed
Inventory decreases in stored feed			Feeding out costs
Grain, nuts and meals			
Allocatable plant and machinery costs			
Agistment and grazing lease costs			
Irrigation water			
Husbandry and marketing			
Depreciation (if any)	Shearing	Running costs of milking sheds (machine parts, fuel)	Medicines and disease control
Freight, levies	Treatment for internal and external parasites		Packing and preparation costs
Marketing	Marketing	Veterinary fees and medicines	Marketing
Medicines and veterinary fees	Depreciation on bought flocks	Replacement rearing	Freight, levies
Interest on bought-in cattle	Medicines and veterinary fees	Bull or artificial insemination	Flock replacement costs (if any)
	Interest on sheep purchased	Herd testing	Interest on livestock bought
		Depreciation of bought-in herds	
		Freight, levies marketing costs	
		Interest on cattle bought	

A system using bought-in female replacements can have as costs (i) deaths, (ii) sire or semen costs and (iii) depreciation. In a self-replacing production system the costs are (i) deaths, (ii) sire and semen costs and (iii) the loss of potential production from alternative animals while the young female replacements are growing to breeding age and weight.

A checklist of main animal costs

Next, in Table 12.1 the main cost items in animal production systems are listed. They can be used to calculate enterprise gross margins (see Chapter 8).

Economic analysis of animal enterprises

In this section the main points regarding costs and returns are discussed. To demonstrate some economic principles which underlie all animal enterprises let us first confine our discussion to a 'simple' animal system – cattle fattening in a feed lot. Two elements – a conversion mechanism and feed – are involved.

Costs

The costs in a feeder situation, where cattle are brought in, fattened and sold are:

(i)   overhead or fixed costs;

(ii) feed costs (including interest on outlays for feed);

(iii) running costs of plant, machinery and casual labour used in feeding;

(iv) husbandry and disease control (this is usually a constant charge per head for given levels of stocking rate and animal throughput, increasing as both the stocking rate and throughput rise, and includes veterinary expenses, medicines and vaccines);

(v) marketing charges (freight, commission, government levies, and saleyard costs);

(vi) interest on purchase price of animals over period held.

Note: losses through death can be accounted for when calculating income.

These costs can be simplified into three groups:

overheads (i);

feed ((ii) and (iii), above);

husbandry and marketing ((iv), (v) and (vi), above).

For a given feed cost, husbandry and marketing charges tend to remain fairly constant per head. Their yearly total depends on the number of animals put through the feed lot, so they can be regarded as variable costs, i.e. costs which vary as throughput rises or falls.

Income

To calculate the profit margin from feeding steers, the initial purchase price of the stock is deducted from the gross income from sale of the fattened stock. The difference also includes loss through deaths. For the purposes of economic analysis, it is convenient to subtract from this net gain the husbandry and marketing costs. We will call the figure so obtained 'the net value of animal product less husbandry and marketing charges' or 'net animal revenue'. The gross margin is calculated by subtracting feed cost from the 'net animal revenue' figure. By using this modified expression of income or gain from the feeding operation, we are able to broadly analyse any feeding operation as a one-input, one-output system. The overheads, since they remain constant, do not affect the point of maximum profitability, which is reached when extra (marginal) costs and extra returns are equal (see Chapter 4).

Modifying the income picture by deducting husbandry and marketing costs means that there are only three components for analysis, viz: overhead costs, feed costs, and 'net animal revenue'. The situation can be expressed graphically (see Fig. 12.1).

The output of the animal, and of the total animal

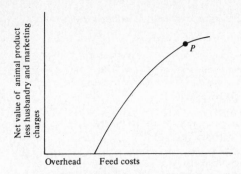

Fig. 12.1. Net animal revenue; the overhead cost does not affect the point of maximum profitability (*P*).

system, is controlled by the level of feed input. As the level of feed increases, the marginal return eventually decreases, as would be expected from the principle of diminishing returns. A stage is reached (*P*), where the cost of an additional unit of feed would be only slightly less than the value of the net animal product which resulted from it. This is the point of maximum profit, for a fixed level of overheads, i.e. when the

$$\frac{\text{extra return}}{\text{extra cost}}, \text{ or } \frac{\text{ER}}{\text{EC}}$$

ratio is approximately 1. Probably no livestock feeder, in practice, is in a position to organise his feed inputs to such a delicate stage, so that the ER/EC ratio is 1. He is more likely to stop the feeding–fattening operation at an input level considerably before this.

All livestock operations can be analysed in terms of the three components listed above, viz: overheads, feed and 'net animal product less husbandry and marketing'. Once we leave the 'factory' type environment of the feed lot, the piggery and the poultry farm, where the feed costs are readily identified, there arises the problem of accurately defining the 'feed cost' figure for items such as improved pasture and hay.

Direct feed costs are generally fairly easy to calculate. In most livestock enterprises they include the cost of purchasing grazing rights, the costs of supplementary feed, and of growing and conserving pasture and/or fodder crops. In irrigation areas devoted to animal production, the cost of water, either bought direct from the irrigation authority or supplied from the farm's own pump and engine, is part of the cost of producing feed for the animals and so can be viewed as a feed cost.

Table 12.2. *An example of the relationship of stocking rate to gross margin*

Goats per hectare	GM per hectare	Extra GM per hectare
3	42	
		8
4.5	50	
		5
6.0	55	
		2
7.5	57	
		−1
9.0	56	
		−3
10.5	53	

With the exception of grossly understocked forage, the amount of feed supplied will determine the level of output from a given grazing-based system of feed conversion. The system of grazing management using the same class of animal can vary greatly.

Once the conversion system begins to operate at reasonable levels of technical efficiency, feed input becomes the main determinant of output. By isolating the role of feed on output, the farm adviser is in a position to use the principle of diminishing returns in determining the level to which he should advise the farmer how much to spend on feed.

Utilising productive pastures and fodder crops

In the following section some economic and management aspects of animal production from grazing good pastures are detailed. Points about feed management which are made below have some relevance to any animal-production system based on (i) the animals harvesting their own feed from forage growing on land over which the animal manager has some control; or (ii) if cut-and-carry is the method of supplying feed; or (iii) if even only a relatively small piece of land and a small number of animals are involved.

The economic value of the unit of feed varies according to the season of the year, because it has a range of effects on the total gross margin. In the same way, the feed produced by irrigation has different effects on total gross margin, as the use to which it is put will give a wide range of extra returns.

Within limits, there is a direct relation between stocking rate and gross margin per hectare on good pasture or fodder crop. Although the highest stocking rate may give the best utilisation, the operation of the law of diminishing returns means that the point of

maximum profit is reached before the point of maximum stocking rate. Table 12.2 illustrates the point. The analysis takes account of the cost of feeding in poor seasons, and loss of revenue from reduction in performance as stocking rates rise.

Biological and financial disasters can result from hasty application of the fact that stocking rate and utilisation are correlated, if the pasture supply is insufficient to support the stocking rates imposed. Vigorous, high producing pastures are a prerequisite of high stocking rates. In many tropical countries, of course, there is not an abundance of vigorous pastures!

On forage-based enterprises in areas where pasture improvement is possible, there are a number of alternative methods of feed supply, and the problem of supplying the greatest amount at the lowest cost requires the application of both technology and economic principles. For example, the farmer may have to decide how much to spend on pasture improvement, where additional inputs can often result in doubling the yearly dry matter production; nitrogenous fertiliser to produce extra growth over a 6–8 week limiting period, e.g. on irrigated pastures; fodder conservation; growing fodder crops; buying feed to help bridge feed shortages; how many hectares to irrigate; and how much water per hectare. The choice of which course, or courses, to follow will largely be determined by:

the period(s) when feed is available from the input;

the quantity of feed supplied in each period (e.g. one programme may produce 3000 kg of dry matter, whereas another, because of restraints, may produce only 500 kg);

the cost per unit of seasonal feed;

the degree of certainty with which it can be grown;

the value, in terms of extra income, of a unit of extra feed available in each period (i.e. the extra return or extra value product of an extra unit of feed in each period). Thus one unit of feed in the dry season may permit an extra animal to be carried for the rest of the year, whereas one unit of feed in the wet season may only allow extra stock to be carried for a short time, so the extra return due to it will be small.

Matching feed demand and supply

The first task in planning livestock activities is to calculate the likely level of feed supply during each

Table 12.3. *Tropical livestock units (TLU)*

Species	TLU	Species	TLU
Camel	1	Horses	0.8
Non-draught cattle	0.7	Mules	0.7
Sheep	0.1	Donkeys (asses)	0.5
Goats	0.1		

season of the year (from each pasture and crop type) and the maximum stocking rate with different animal activities using data on the feed requirements of livestock.

The major problem in planning livestock enterprises in which the animals graze improved pastures and crops is to specify seasonal feed supply and carrying capacity. Although grazing and cutting trials are an important source of information, the experience of observant managers of individual farms is still the most important source of basic data for planning systems of livestock feeding and management.

One very crude unit of measure of energy demand of grazing ruminants (cattle, camels, sheep and goats) which is also sometimes applied to equines, is the Tropical Livestock Unit (TLU). It is calculated on the energy needs of a 250 kg ruminant. The above rankings apply (Table 12.3).

For accurate planning, the feed requirements of animals at various stages of growth, pregnancy and lactation have to be matched with feed supply from existing pastures, crops, conserved and purchased feedstuffs.

The best way to calculate animal demand is to break the 'animal year' into a number of sub-periods, e.g. dry, pregnant, lactating, growing, post-weaning, fattening. An attempt can then be made to match needs with the expected seasonal feed supply. P. A. Rickards and A. L. Passmore in their publication *Planning for Profit in Livestock Grazing Systems*, published by the Agricultural Business Research Institute (ABRI), University of New England, Australia in 1983, have come to grips with this problem. Their basic unit is the Livestock Unit (LSU), which is a 45 kg non-pregnant, non-lactating sheep, grazing reasonable pastures, with an energy exercise factor of 35% above basic metabolic needs.

Table 12.4. *LSM ratings for stock at different stages of production cycle*

Weight (kg)	Type	Performance	Monthly energy needs (LSM)
45	Sheep	Maintenance	1.0
250	Steer	Maintenance	5.5
250	Steer	Gaining 0.68 kg/day	7.7
250	Steer	Losing 0.45 kg/day	4.8
360	Cow	Last month of pregnancy	10.1
360	Cow	Producing 9 l/d of 4% fat milk	12.0
300	Horse	Working in field	8.3
300	Horse	Not working	6.2

The energy needs of this LSU are expressed in units of kilocalories of metabolisable energy (ME). ME is the international measure of animal energy usage. In this case, it is 60 000 kilocalories, i.e. (60 Mcal) per day. Note: 1 Mcal = 4.2 Mj, which is the current metric measure of energy. The energy needs of one LSU for one month is known as a Livestock Month (LSM). Thus, monthly maintenance energy requirements for a 250 kg beef steer are 5.5 LSM. Requirements for pregnancy, growth and lactation are also expressed in LSMs. Some examples are shown in Table 12.4.

Expressing feed supply in LSMs

The energy supplying power of different pastures, crops and concentrate feeds can also be expressed in LSMs, i.e. how many months one LSU can be fed for, on different classes of feed. Thus the LSM unit is a link between supply of energy from feed and demand for energy from feed, which allows the planner to match the known demand of animals with expected supply of feedstuffs. In Table 12.5 we show the LSM values for different feeds.

Intensive production feeding

The most crucial aspects of any animal farm management is the management of feed. That is, to ensure that a sufficient quantity of feed of suitable quality is available so that the animals can fulfil their productive potential.

Table 12.5. *LSM per tonne of various feeds*

Type of feed	Dry matter (%)	LSM
Medium quality pasture or legume hay	85	30
Good green pasture	25	8.6
Poor green pasture	30	6.0
Green fodder sorghum	25	8.8
Crushed dry maize	88	48.4
Crushed dry sorghum	88	44.0

Usually much closer attention is paid to the cost and quality per unit of feed in intensive animal-producing enterprises, i.e. poultry, pigs, beef feed lots, than in enterprises based largely on grazing of pastures and crops. The two main reasons for this closer scrutiny are that most of the feed is bought, and that the possible low margin between feed cost and prices received for the animal product means that an unwise feed purchase can result in big losses. The keys to obtaining profits in any intensive animal system are:

low cost, good quality feed;

proper attention to disease control and husbandry, and minimising the animals' psychic stress – neglect of these three factors in the intensive environment leads quickly to disaster (it is like sitting on a volcano)!

Draught animals

In many areas draught animals will become, or will continue to be, the chief source of power. This will be because of their suitability to the task, and operational and economic advantages, as compared to either tractor and/or human power. The adequacy of animal power to carry out tasks will be enhanced by implements and harnessing being better designed and made from improved materials. Draught power can be a cheap source of power which also provides meat, milk, saleable stock and, importantly, manure. It is also self-replacing. The working capacity of a draught animal is determined by the kind and size of the animal, the method of harnessing, the speed of travel, its training, its health, and (the key to it all) body condition and nutrition – especially energy intake.

This raises the dilemma of where the feed for the well-fed working animal comes from. Good farmers who rely on draught animals recognise their true economic contribution to the farm operation. Consequently, they take special pains to supply them with good rations. Measures used include growing a fodder crop on a special plot, giving them first priority in grazing new crop residues, keeping the better quality household wastes for them, retaining some of the surplus grain available for sale and even obtaining feed from outside the farm through purchase or exchange. The condition of the work animals tells a lot about a farmer.

For profitable animal production, be it on grass or factory farm, it is vital that the operator have a good grasp of current knowledge concerning different production systems, replacement systems, feeding costs, husbandry, animal-feed demands, pasture-feed supply, and importantly, matching feed demand and supply.

Breeding

In discussing genetics it is important never to lose sight of the fact that a large proportion of the variation in production characters observed in farm animals is due to environmental factors, i.e. animal feeding and management has more influence on production than does the manipulation of animal genes.

The contribution which animal geneticists have made to increasing animal production in most of the tropics has not been spectacular. In contrast, plant geneticists have had a much greater impact. Not enough work has been done to improve locally adapted animal species. We noted in Chapter 3 how importing cattle from temperate areas into developing tropical countries had often led to great disappointment. These cattle were not adapted to the environment. As a result, they either died, or dropped their production levels to those of the local strains or worse.

So, it is important to distinguish genetic from environmental effects. The adage 'half the breeding goes down the throat', sums up past experience with breeding programmes. It should help people to make decisions on how best to capitalise on the findings of research into the applied aspects of population genetics. There are two ways of breeding livestock with superior production characteristics:

(i) selecting within a given population or breed;

(ii) cross-breeding.

The second method gives the quickest gains, and is the method chosen not only by farmers who have relatively few animals, but also by many owners of large flocks or herds who find they can make a good deal more money by cross-breeding the poorer one-third of their females, than by straight breeding. First we will consider how to apply the principles of population genetics to improving genetic worth within a given breed.

Population Genetics

The basis of selection of superior characteristics in animals of a given breed has changed from an emphasis on the individual to that of the population. Within any animal population there is a variation in the level of individual performance expressed (for example, different milk production, growth rate, adult size, carcass composition, kidding rate).

Animal breeders can use this variation to gain genetic improvement by keeping, for future breeding, those animals which express the highest levels of the desirable traits. However, only part of the better performance of the 'high' group within this population of animals is due to a superior collection of genes. The rest of the variation is due to the effects which different environmental conditions have on animal performance.

In any breeding programme the geneticist tries to isolate the relative contribution of genes and environment. Geneticists have introduced the concept of heritability. A character is highly heritable when the environmental effects on variation in a production character are small relative to genetic effects, and vice versa. An example of the meaning of heritability can be seen when selecting cattle for weight gain at 400 days of age (this is a more valid selection criterion than weaning weight). Let us assume a breed which averages 300 kg at 400 days, with a range from 200 to 400 kg. After correcting for sex and minor age differences, the bulls of the 400 kg group are kept for breeding and are later mated with a group of females which averaged 300 kg. The average 400 day weight of their progeny, if

Table 12.6. *Heritabilities of production characteristics of various classes of livestock*

Animal	Characteristic	Heritability
Merino Sheep	Greasy wool weight	0.3–0.4
	Twinning	0.32
	Milk yield	0.30
Beef Cattle	Birth weight	0.40
	Weaning weight	0.30
	Post weaning weight gain	0.50
Dairy Cattle	Milk production	0.25
	Fat	0.55
	Body size	0.40
	Mastitis resistance	0.25
	Ketosis resistance	0.25
	Milking time	0.40
Poultry	Body weight	0.40
	Egg weight	0.60
	Shell colour	0.60
	Age at first egg laying	0.50
Pigs	Pigs farrowed	0.15
	Pigs weaned	0.10
	Feed conversion	0.30
Horse	Racing ability	0.45
Mouse	Tail length	0.60

grown under comparable conditions, will be 320 kg, not 350, as you might expect.

This is explained as follows: the heritability of 400 day weight gain is about 40% or 0.4. Thus, of the observed 100 kg superiority of the selected adults, only 40% or 40 kg is heritable and so passed on. Since the females joined have no relative genetic superiority, they can contribute nothing to increasing the weight of their progeny.

Males	Progeny	Females
+ 100 kg		0.0 kg
× Heritability 0.40		0.40
Contribution 40 kg →	$\frac{40+0}{2}$	← 0.0 kg
	= 20 kg	

Average 400 day weight of
progeny = 300 + 20 = 320 kg

The heritability figures shown in Table 12.6 above have been derived from large-scale experiments and real-life observations. They are not 'theoretical'.

If the animal breeder is trying to select for two characteristics occurring independently of each other,

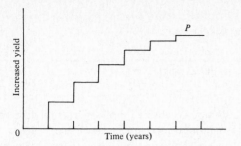

Fig. 12.2. The increase in yield of progeny due to using sires with superiority in the chosen characteristic. *P* is the sires' potential, which sets an upper limit.

e.g. carcass weight and reproduction rate, the rate of genetic progress will be halved. The more independent characteristics selected for at a time, the slower the progress towards the overall objective. Thus, if some characteristics have to be dropped from the programme, they should be the ones of least economic importance. Generally, characteristics affecting reproduction have low heritabilities but high economic importance, while carcass characteristics have both high heritability and economic merit.

For most species, one male can fertilise at least 40 females. So it is more difficult in practice to obtain a population of females with as great a degree of superiority relative to the average male, i.e. the selection pressure the breeder can exert is much higher for males than for females. The old saying 'the sire is half the herd' conveys the same point more vividly.

Economics and genetics

The genetic gains which are theoretically possible, and the economic values which can be attached to such gains, are strongly modified by each particular farm environment. Specific farm characteristics and management strategies determine the genetic gains and economic rewards which can be had. A diminishing-returns effect occurs in a long-term breeding programme because the degree of superiority decreases each year with increases in the character being selected for in the progeny. If sires or semen used in artificial insemination of the same quality are always used in a breeding programme, the sires' contribution each year would be two-fold:

(i) they will hold the gains already made above the base level;

(ii) they will add to the existing gains, another gain of slightly less magnitude than the gain made the year before.

If the programme continues long enough, the performance of the whole population will gradually approach that of the (age and sex) corrected performance of the sires, provided that the environment of the stud, and the herd or flock under consideration, are broadly similar. The graph of progress in Fig. 12.2 has three features:

(i) there are annual increases in production of the selected characteristics;

(ii) the increase occurs at a diminishing rate over time;

(iii) for all practical purposes, performance beyond point *P*, which is close to the sires' performance, can be treated as constant.

The step line can be converted to a curve, and the performance trait converted to income (Fig. 12.3).

Note: a major practical point emerges if you think about these figures, viz: when applying genetic principles in the real world, it is hard to make progress, but conversely, it is just as hard to go backwards completely.

One economic analysis which can be done is to convert the expected future income from the genetically induced increased production, to today's dollars, and then compare these gains with those which can be achieved in other ways, e.g. through improved housing or better feeding.

Another short-term analysis can be made by simply putting dollar values on the immediate (next generation) gains. This quick analysis gives a rough guide to

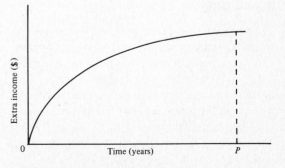

Fig. 12.3. The increase in income due to using superior sires.

the sort of maximum premium which could be paid for the superior sires or semen for artificial insemination. However, the maximum premium which can be justified depends on genetic gains which are made (the benefits of which are hopefully maintained into the future). A direct economic 'gain' of, say, $50 in the next generation from using superior sires or semen is not the total gain, as the improvement contributed by the sire is maintained into the future by his progeny. It is becoming increasingly accepted that action to improve animal production needs to be directed to the production characteristics which are of greatest economic importance.

One practical application of the principles of genetic gain is animal selection based on measured performance. The benefits of herd improvement in dairying by selection based on production figures and the widespread availability through artificial insemination (AI) of performance-tested superior sires, are well known.

One common difficulty is finding the so-called 'superior' animals we have been discussing. Where flock or herd numbers are small, use of AI and participating in group breeding schemes are likely to be worthwhile. This way the benefits from applying the 'law of large numbers' in genetic improvement can be gained. For example, in Australia, thousands of farmers belong to a group breeding scheme for Merino sheep. About 1 million female sheep are involved. The best 1–2% of each person's flock goes to a central, elite pool, and then the rams which are bred from this pool go back to the individual members of the society. Now that the technology of storing ram semen has been developed, the use of AI will further help speed the rate of genetic progress.

Cross-breeding

The benefits of cross-breeding are well established and widely exploited in the pig and poultry industries. Cross-breeding is not so widespread in the cattle industries. Still, cross-breeding between basically similar cattle breeds can increase production by 20% or more compared with pure breeds. Similar benefits have been reported for other species (sheep, chickens, pigs). The hybrid vigour (heterosis) effect of cross-breeding, in the case of cattle, shows itself in higher:

calving rate;

calf survival;

weaning percentage;

weaning weight;

post-weaning gain.

Whilst the advantage of the cross for each one of these production characters may be only 3–4%, their effect is additive; i.e. more calves are born, these calves survive better, more of them are weaned and at a higher weight, and after weaning, they grow better. Thus the cumulative effect of these advantages is to give a production increase of up to 20%, which is a much bigger and far quicker gain than can be obtained by selecting within a breed. No great advantages in carcass quality have been observed from cross-breeding. Sometimes there is also scope for introducing new breeds or new species profitably into an existing system.

Questions

1 How important is animal production in the local farming? Is there potential for increasing animal production within the existing farming systems?

2 What are the main factors which control the efficiency of the animal as a feed-conversion mechanism?

3. What are the major limitations on animal production in your area?

4 What are the basic techniques for analysing the economics of any animal system?

5 What do we mean when we say that feed has different economic value at different times of the year?

6 What are the commonly used standard measurements of the energy requirements of animals? How do you use these in matching the feed supply and demand of pastures and animals?

7 What are the main factors in operating a successful intensive production unit?

8 What is meant by the following statements on animal breeding:
   'half the breeding goes down the throat';
   'the sire is half the herd'?

13

Mechanisation

Theory

The term 'mechanisation' has a wide meaning. It embraces the use of items such as hoes, sickles, ox and buffalo ploughs, threshing and winnowing machines operated either by hand, foot or animal, pumps, electricity-generating plants and milking machines. All of these 'machines' have some common features:

They wear out They have to be replaced, after a time, if the farmer wants to continue to use the services which they provide.

They need maintaining and repairing The hand cultivator needs to keep his hoe sharp, the owner of a tractor needs to maintain it (using grease and oil, or by adjustment) and to repair broken or worn-out parts.

They use energy This can come from:
the food humans eat, which provides energy for human activities;

animal power;

wood, straw and wastes;

fossil fuels (petrol, oil, coal);

sunlight (solar power);

electricity.

They provide services These include preparing seed-beds, pumping, threshing and can be obtained in several ways (these include owning, hiring, contracting, leasing, share farming and exchange).

The key economic question facing the farmer is how to decide which form of machinery service can he obtain at lowest cost.

Details of machine costs

There are a number of types of costs involved in any machinery operation. For example, some of the costs of owning and operating a tractor are: fuel, labour, tyres, lubrication, repairs, depreciation, insurance and housing. Specific costs are discussed below.

Variable costs

Costs such as fuel, labour, tyres, lubrication, repairs, are 'variable' or 'direct' cost, i.e. they remain fairly constant per hour of operation and, consequently, vary mostly with the hours of operation. Small variations occur in:

fuel used per hour, which varies according to engine load, size and age of the engine;

labour costs per hour of tractor operation because the ratio of man hours worked to tractor-operated hours is different with each type of operation (on average this can be taken as 1.3 man hours per tractor hour operated);

repairs per hour, which generally increase as the hours accumulate (a proportion of repairs is related to the time that the machinery is used as well as to how often it is used).

In general, these costs (other than repair costs) can be considered to occur at a constant rate per hour.

Overhead or 'fixed' costs

Another class of machine costs are classified as overhead or 'fixed' costs. Some of these costs are 'hidden'. The main overhead cost is depreciation. The machine wears out each year and there is a cost of replacing it (maybe in several years' time). This is called

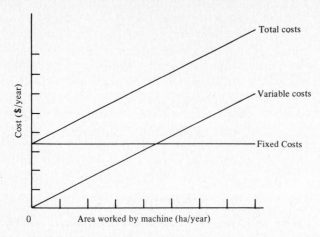

Fig. 13.1. Total, fixed and variable machinery costs/year.

depreciation. Total depreciation is made up of time depreciation (the loss in value due to the passing of time) and use depreciation (the loss in value because of long hours worked). For example, two identical tractors of the same age would have different values if one had done 1000 h of work and the other 5000 hours. Obsolescence is the unpredictable loss in value which occurs when a greatly improved model is released. It is a form of time depreciation. Total (use and time) depreciation is usually higher during the early years of a machine's life. (Theoretically, the loss in value or depreciation is deducted from operating profits each year and invested until the machine needs to be replaced.)

Sometimes it is difficult to know whether to class a cost as depreciation or as repairs. One rough way of avoiding this difficulty is to treat all depreciation as an overhead cost and all repairs as an operating cost. The overestimate of depreciation will be partially offset if all repairs are treated as operating costs.

Overhead costs can also include the costs of keeping a machine in a shed. If it is not stored in a shed then these costs occur anyway, in the form of increased depreciation and repair costs due to the machine being left out exposed to the weather.

Timeliness costs
If the machinery service is not performed on time, there usually will be a loss of income. For instance, the effects of delays in preparing the seed-bed, weeding,

harvesting, threshing and milking are well known to most farmers.

Opportunity costs
On occasion it is appropriate also to count the opportunity cost of the money tied up in the machinery. This can be the interest cost of the annual average value of the capital tied up in the machine over its life. The rate of interest to charge as opportunity cost is that rate which the capital could be earning if put to the best alternative use for it. The question of whether real or nominal interest rates should be used has to be resolved; the answer depends on whether the sums are in real or nominal dollars (see Chapter 9). We use real (today's) dollar terms and real interest costs when comparing alternatives, and then, once a decision is made, we redo the sums in nominal dollars to find out how many nominal dollars the decision-maker will have to set aside for depreciation and replacement, or to ask the bank for to implement his plan.

Standard of operation
The standard of performance of a piece of machinery can be a 'hidden' cost to be considered in the decision on which form of machinery service to use. For example, more or less grain may be obtained with a reaping machine than by hand reaping with a sickle. Perhaps an ox or buffalo plough does not prepare as good a seed-bed as a small tractor does. Mechanical threshers may get a higher percentage of grain from the

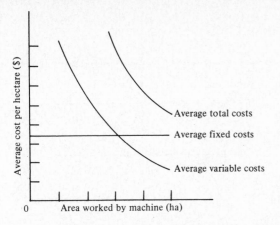

Fig. 13.2. Average total, fixed and variable machine costs/ha.

reaped crop than do hand beating or buffalo trampling. Further on in this chapter (pp. 133–7), we show how to take account of all these different classes of cost. The example that we use deals with getting the services of a tractor.

Annual usage

In Figs 13.1 and 13.2 the way in which fixed or variable costs contribute to total annual costs and total cost/ha are shown.

The less a machine is used, the higher the fixed cost per unit (per hectare, per hour). As the machine is used more, the fixed costs are spread, which reduces the unit costs. The annual hours of use or the annual output of a machine will determine the relative unit costs of a large or small machine. For example, if a larger tractor has an annual overhead cost of $4000, while that of a smaller one is $2000, and if the small tractor works 200 h/year to complete 200 ha of cultivation, its overhead cost is $10/ha. The larger tractor can work 200 ha in 100 h, giving an overhead cost of $20/ha. However, if it works for 500 h and completes 1000 ha of cultivation, the overhead cost will drop to $4/ha. In practice, about 600 tractor hours per annum are needed to achieve reasonable economies in overhead costs.

Costs of owning and contracting machinery services

Costs per hectare or hour of use can be used when comparing the alternatives of owning a machine or getting a contractor to provide the machine service which is required. The costs of owning or contracting are compared in Fig. 13.3. Also, the break-even number of hectares (the minimum number of hectares to justify owning the machine) per year is indicated.

Obviously, it will not be economic to own very expensive machinery which will have little annual use. Below a certain level of machinery use it will be cheaper to pay a contractor to carry out the task. Conversely, above a certain level of machinery use it will be cheaper to own the machine than it would be to hire contract services.

Some problems in estimating the cost of machinery

In figuring the economics of owning a machine before making a decision about alternative machines, the farm adviser must decide which costs are to be taken into account. Generally, only the extra costs are considered when advising about committing money to machinery. The approach is essentially that of partial budgeting. Extra costs for machines are those which occur directly as a result of purchase and/or operation of the machine. Costs which occur even if the machine is not purchased are not included; also if the farmer already owns a machine and wants to use it on another enterprise, the only costs to be considered are the extra operating costs. Where ample shed room is already available, housing costs are omitted from the calculations.

Labour costs in total machinery costs

It is vital to know the time taken for various machine operations (i.e. work rates) so that, if a change in ways of performing tasks is being considered, reliable data are available.

The rewards to the farmer, as a manager/owner of a farm business, come as his 'normal' living costs plus any after-tax farm 'profit' left over. If casual labour is employed for the operation of a machine, labour costs are charged at the direct wage costs. Where permanent labour is employed, the workers are normally paid a weekly or annual wage, regardless of whether they are operating machinery, repairing it, or doing any other jobs during slack times.

We aim to be realistic, not doctrinaire. We suggest that once a decision has been made about the form of machinery and of labour force, then no charge be made

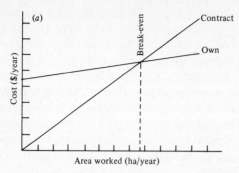

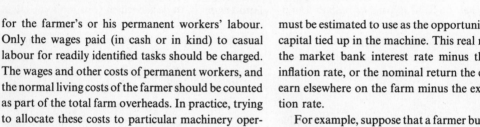

Fig. 13.3. (*a*) Breakeven costs and areas for machinery ownership versus contract services; break-even number ha/year between ownership and contract of machinery services are illustrated. (*b*) The change in costs/ha between ownership and contracting, and the break-even number of hectares.

for the farmer's or his permanent workers' labour. Only the wages paid (in cash or in kind) to casual labour for readily identified tasks should be charged. The wages and other costs of permanent workers, and the normal living costs of the farmer should be counted as part of the total farm overheads. In practice, trying to allocate these costs to particular machinery operations or other farm activities is usually a waste of time.

Interest costs in total machinery cost
When comparing the costs and profits from using alternative sources of machinery services to do a particular job, some of which involves capital investment and some of which does not, it is necessary to charge interest on the capital involved, if the comparison is to be valid.

The conventional way in which interest is treated in machinery costing is to make a charge for interest, to account for the cost of using capital in this particular way. To mean anything, the interest charge needs to be based on the average value of the machine over its expected lifetime. We use the average of the purchase price and trade-in value of the machinery in constant (today's) dollar terms as a rough guide to the average amount of capital invested in the machine in any one year of its life. This average value is the sum of the purchase price and trade-in price, divided by 2. As the calculation is done in constant (real) dollar terms then the interest rate to use must also be in real terms (see Chapter 9 for a full explanation). So, a real interest rate

must be estimated to use as the opportunity cost of the capital tied up in the machine. This real rate could be the market bank interest rate minus the estimated inflation rate, or the nominal return the capital could earn elsewhere on the farm minus the expected inflation rate.

For example, suppose that a farmer buys a machine for $6000 and expects to trade it in after using it for 5 years, i.e. at the start of Year 6. It will then be worth about $1000. The machine will have depreciated by $1000 for each year of use. The value of the capital tied up in the machine will then be:

$6000 at start of Year 1;

$5000 at start of Year 2;

$4000 at start of Year 3;

$3000 at start of Year 4;

$2000 at start of Year 5;

$1000 trade in value at start of Year 6.

The average value of the capital tied up in the machine is $6000 + $5000 + $4000 + $3000 + $2000 + $1000 = $21 000 ÷ 6 years
Average value = $3500

A simpler way to work out the average value is to add the purchase price and trade in value, and to divide by 2:

$$\frac{\$6000 + \$1000}{2} = \frac{\$7000}{2} = \$3500$$

The 'opportunity' interest cost charged against this average value (real) of capital then needs to be a real, not a nominal, interest rate.

The idea of using average value, rather than purchase price, is that the amount deducted from income each year as depreciation, is supposedly invested, and therefore earns interest. This partly offsets the interest charged against the original purchase price. The interest earned on the money set aside from depreciation does not *fully* cover the interest charge on the purchase price. If interest is charged on the average value of the machine, the resultant figure is good enough for most practical purposes.

We have pointed out several times that the correct interest rate to charge when comparing alternatives should be the opportunity cost of the capital invested. Using the bank overdraft figure is common, but is often too crude for real-life situations involving many alternative uses of capital.

When not to charge interest

Where a farmer has gone through the process of choosing between alternatives already and has bought the machinery, then no interest should be charged when calculating the operating costs of the machinery.

Certainly, under the general item of the whole farm's finance costs, it is essential to provide for the actual interest and loan repayments that the individual farmer has to make to repay the cost of the machinery. This is not, however, part of the actual operating costs of the machine, which is the figure used to calculate the costs of the farm activities.

The essential point is that the situation of the individual farmer will govern how much interest and principal he has to pay. For example, some pay cash for a piece of machinery and so have no repayments. Others can make up only a one-quarter deposit, and so have large annual repayments. Then there is the whole range of situations between these extremes. So, it is not possible to give a general prescription on what values to use.

Depreciation and replacement

Machinery prices (and running costs) can increase rapidly due, in part, to high rates of inflation. We discussed the general problem of inflation in Chapter 9. Now we apply the principles outlined there to the following problem. How much money should be set aside* each year to cover both depreciation and inflation, so that there will be enough money available to replace the machinery when it is either worn out, or due for replacement?

In tackling this problem it is very important to be very clear about the dollar terms we are using – real or nominal. As we explain in Chapter 9, an easy trap to fall for is to mix real and nominal dollars in the same calculation, giving some hybrid results.

In calculations to compare machinery alternatives we have found that it is simplest to do all the sums in real dollars. This presents no great problem in estimating variable costs, other overheads, or purchase price of a machine, as we use today's figures. However, some problems are encountered when estimating the amount of money which needs to be set aside each year to allow for the cost of the machine wearing out, and the need to replace the machine when it does wear out at some time in the future (i.e. future figures will most likely be affected by inflation). In taking account of the cost of replacing a machine in the future it is necessary to take inflation into account, so that the sum that is put away each year adds up to enough inflated dollars to replace the machine.

We will show in the following example how to estimate the replacement sum to use in calculating the annual cost of owning a machine. To compare alternatives, we will do all the sums in real dollars. Then, to know how many actual dollars to put aside in each year, we use nominal dollars. This distinction is very important.

Comparing alternatives: real terms Suppose that a machine is worth $1600 today and is expected to have a working life of 8 years. It will have some trade-in value at the end of its useful life. It is reasonable to expect that it would be worth $400 (of today's dollars) in 8 years' time. We can say that the farmer will need $1600 minus $400 ($1200) of today's dollars in 8 years'

*In practice, farmers usually do not put this money in the bank, because it can be used more fruitfully by employing it to produce income on the farm. Even so, it is necessary to provide for the cost of replacing the machine. Our method, though over-simplifying the situation, helps meet this need.

time in order to replace the machine, i.e. $150 per year is the real depreciation-replacement cost (8 × $150). If the farmer were to put $150 away each year, buried in a jar near his house, then in 8 years' time when he needs to buy the machine he will have the funds ($1200) saved up.

However, there are better ways of accumulating the required $1200. Instead of putting the money in a jar each year (where no interest is earned), the owner can invest some money each year at compound interest. He can invest an amount of money each year which, provided that he leaves it in the bank, will grow at compound interest to $1200 in 8 years' time.

His adviser can calculate how much money the farmer ought to set aside each year in order to save up the $1200 which will be needed to replace the old machine. He looks up the sinking fund factor (annuity whose terminal value is 1, Table E in Appendix 1). For an 8 year time period, at a 2% real interest rate, the sinking fund factor is 0.1165.

This means that if he invests 0.1165$/year for every one of 8 years, at 2% real compound interest (i.e. he reinvests the interest each year), he would have $1 in the bank at the end of 8 years.

Amount required in 8 years time	Sinking fund factor	Amount to invest each year at 2% compounding to give $1200 in 8 years time
$1200	× 0.1165	= $140

So, if he invests $140 each year he will have $1200 by year 8. As a check on this calculation, from Table D in Appendix 1 (terminal value of an annuity) it can be seen that $140 invested each year at 2% compounding interest grows by a factor of 8.58 in 8 years' time to $1200. Thus, the $140 is the real annual cost of depreciation. It is the figure to be used when comparing the costs of alternative machines, or alternative forms of machinery service.

Depreciation and replacement cost: allowing for inflation. Once a decision is made, the farm adviser must take inflation into account to know how many inflated dollars the farmer will need to put aside in each year to be able to replace the machine. If inflation is expected to be 9% per year over the life of the machine then the cost of the new machine will be as below:

Purchase price of machine today	Multiplied by compounding factor of 9%	Cost of machine in 8 years, i.e. amount needed to replace it
$1600	× 1.99	= $3184 (say $3200)

The trade-in value of $400 will also inflate, but probably not at the same rate as the general inflation rate. So, say that the trade in value inflates at 5% per year.

Trade-in value today	Compounding factor of 5%	Trade-in value in 8 years
$400	× 1.47	= $588 (say $600)

Amount needed to replace machine in 8 years' time:

Replacement cost of machine	$3200
Minus trade-in from old machine	$600
Amount needed to buy new machine	$2600

Instead of putting $325 away each year somewhere, and not earning interest ($325 × 8 = $2600), the farmer can invest a sum each year at the market interest rate of 11% (assuming 9% inflation, 2% real, compounding). As before, find the sinking fund factor to accumulate the $2600 in 8 years at 11% interest, i.e. 0.0843.

Amount required in 8 years time	Sinking fund factor	Amount to invest each year
$2600	× 0.0843	= $219

Thus, the $219 is an inflation-corrected depreciation cost which must be charged against the annual costs of owning the machine. It is the actual sum to put aside each year, not the $140 we used when comparing alternatives.

The annual ownership cost of this machine is both $140 and $219 nominal; the sum to use depends on the purpose for which the calculations are being done, and the values which are being used for other, related sums.

Practice

Acquiring machine services

The three main ways of getting tractor services are owning, contracting and share farming. Some of the basic elements in owning and operating a medium-sized tractor are presented in Tables 13.1, 13.2 and 13.3.

These figures compare with the $11.10/h for 500 h annual use.

Remember: we have not included labour costs. After about 600 h/year, hourly costs do not drop very much. They rise markedly as annual hourly use falls below 300. This is partly why it pays most small farmers to use contractors rather than owning their own tractor; of course, they also usually don't have enough capital to buy their own machine.

Working Life of Machines

The working life of a machine can be measured in total hours or total areas worked. Often, number of years of use is used in place of both of these measurements. So, to calculate the annual overhead and variable costs of owning a machine for, say, the 'most likely' length of working life (e.g. it may be 5000 h or 10 years) for a particular machine in local circumstances it is useful to work out the costs for the situation where the machine is replaced after different periods of ownership.

Variation around the 'most likely' length of working life is quite possible, even probable. If things are going badly, a couple of extra years of use can almost always be 'squeezed out' of a machine. Similarly, in some circumstances (to do with prices of output, input, inflation, seasons, technical developments, credit for mechanisation) it may be profitable to replace the machine after 5 years. Once the overhead and variable costs of owning and operating the machine are calculated, these figures can then be compared with the costs of other forms of acquiring machinery services. It is then necessary to weigh up how the difference in operating costs of contract and share farming stand up against the expected annual costs of ownership, considering the likelihood of the various lengths of working life.

Table 13.1. *A simplified budget of annual total operating cost of a $20 000 tractor used for 500 h/year (PTO, 45 KW)[a]*

Assumptions	($)
Tractor is sold after 5000 hours (10 years)	
Expected difference between new price and trade-in price in today's values: $20 000 − $5500	14 500
Amount needed to be set aside each year as depreciation, and invested at 2% to grow to $14 500 in 10 years' time, so that the tractor can be replaced: $14 500 × 0.0913, in real dollars	1324
Cost of replacing tractor in 10 years' time with inflation rate for cost of new machinery of 8%/year: $20 000 × 2.1 (compounding factor)	42 000
Present trade-in value of 10 year old, well-maintained tractor with 5000 hours	5500
Expected trade-in price in 10 years' time assuming inflation rate for second-hand machinery to be 5%/year = $5500 × 1.62 (compounding factor)	8910
Expected difference between new price and trade-in price in 10 years' time: $42 000 − $8910, i.e. the nominal amount needed to replace the tractor	33 090
Amount needed to set aside each year as depreciation, and invested at 12% to grow to $33 090 in 10 years' time so that the tractor can be replaced: $33 090 × 0.056 (sinking fund factor), in nominal dollars	1850
Repairs and maintenance, 5% of original new price per year	1000
Fuel (500 h × 12 l/hr × 40¢/l)[b]	2400
Lubrication	100
Tyres and batteries	350
Insurance and road permit	120
Interest on average value of machine, at 2% real interest opportunity cost: $\dfrac{20\,000 + 5500}{2} \times \dfrac{0.02}{1} =$	255

[a] Labour costs are excluded.
[b] A fully loaded diesel engine uses about 3 l/h of fuel for each 10 kW of engine capacity. Average fuel consumption would be less than this, however, as tractors are not fully loaded at all times and, in many cases, are rarely fully loaded.

Table 13.2. *Hourly and yearly costs for $20 000 45 KW tractor, operating 500 h/year*

	Year ($)	Hour ($)
Overhead costs (OH)		
Allowance for depreciation	1324	
Interest (0.02% real)	255	
Insurance	120	
Subtotal (OH)	1699	3.40
Variable costs (VC)		
Repairs and Maintenance	1000	
Fuel	2400	
Lubrication	100	
Tyres and batteries	350	
Subtotal (VC)	3850	7.70
Total costs	5549	11.10

Effect of the length of life of machine on the overhead costs

We will consider two cases:

(i) where the tractor is kept for 5 years (2500 h);

(ii) where it is kept for 15 years (7500 h).

In both situations, inflation on new tractors is assumed to be 8% per year, and on second-hand ones, 5%.

Trade-in prices, in today's dollars, are estimated as 40% of new value for a 5 year old tractor and 15% for one which is 15 years old. Thus trade-in values are:

5 year old = 0.40 × $20 000 = $8000
15 year old = 0.15 × $20 000 = $3000

Case 1: Keeping a tractor for 5 years

Cost of new
replacement in 5 years'
time   = $20 000 × 1.47 (8%
                           inflation factor)

(8% inflation)   = $29 400

Trade-in value in
5 years' time   = $8000 × 1.28 (5%
                           inflation factor)

(5% inflation)   = $10 240

Table 13.3. *Effect of annual hours of usage*[a]

	125 h	250 h	1000 h
Overhead costs	13.60	6.80	1.70
Variable costs	7.70	7.70	7.70
Total costs/h	21.30	14.50	9.40

[a] Costs per hour for 125, 250 and 1000 h/year (real dollars).

Money needed to
replace tractor   = $29 400 − $10 240
                          (replacement price
                          minus trade-in value)
                        = $19 160

Annual depreciation:
replacement cost at a
12% interest rate of
earning for 5 years   = $19 160 × 0.16 (sinking
                          fund factor)
                        = $3065

Using the same method and assumptions, the annual sum needed to be set aside to replace the tractor in 15 years' time is $1537. Interest (12%) on the average value of the tractor, in nominal dollars, is calculated below.

Case 1 (5 years)

$$\frac{20\,000 + 8000 \times 0.12}{2}$$

= $1680

Case 2 ((15 years)

$$\frac{20\,000 + 3000 \times 0.12}{2}$$

= $1380

Assuming that insurance decreases as the tractor ages, the average annual overhead costs for keeping the tractor for various lengths of time are as follows:

	5 years (2500 h) ($)	10 years (5000 h) ($)	15 years (7500 h) ($)
Depreciation and replacement allowance	3065	1850	1537
Interest on capital	1680	1530	1380
Insurance	150	120	80
Total/year	4895	3500	2997
Total/h	9.8	7.0	6.0

Table 13.4. *Cost of machine services for 500 h work*

	Own ($)	Contract ($)	Share ($\frac{1}{3}$ share) ($)
Gross income *(less harvesting costs)*	30 000	27 000[a]	10 000
Costs			
Variable growing costs *(excluding tractors)* (seed, fertiliser, sprays)	12 000	12 000	4000
Tractor costs for cultivating and sowing (400 h)	4440 (400 × 11.1)	7640 (400 × 19.1)	—
Tractor costs for transport (100 h)	1110 (100 × 11.1)	1910 (100 × 19.1)	—
Tractor driver (permanent)	3500	Nil[b]	—
Sub-total of costs	21 050	21 550	
Surplus to pay other costs and profit	8950	5450	6000

[a] 10% less due to untimeliness.
[b] Contractor pays.

The main reason for the drop in overheads as the period of ownership lengthens is that the tractor loses value more quickly earlier in its life than later. Against this, there is often a rise in the variable costs per hour, as the machine ages, especially in repairs and fuel consumption per hour.

Choosing between alternative sources of machinery services

Machinery services can be obtained in various ways (by ownership, contractors, sharefarmers and also from leasing). There are costs – both direct and hidden – associated with each form of service. We will work through an example to show how the relevant factors should be taken into account when deciding on the 'best' or the 'least cost' form of machine service. The reader can apply his own numbers to his situation.

In this example, we have assumed that a medium-sized semi-commercial farmer has 30 ha of arable land which is cropped in a stable rotation of grain, legumes and cereals, in a double-cropping system. He needs tractor services for ploughing, sowing and transport. Harvesting is done by hand with casual labour. A

minimum of 10 h tractor time per hectare (total: 300 h) is needed each year for cultivation and sowing plus another 100 h for transport. To cope with the inevitable problems and emergencies, another 100 h of contingency (or reserve) tractor services is added to the estimated minimum of 400 h. So, 500 h of tractor time are needed.

His choice is between buying his own tractor, or using either contractors or sharefarmers for the cultivation, sowing and transport services. With sharefarmers, he gets $\frac{1}{3}$ of the gross income, pays $\frac{1}{3}$ of seed, fertiliser and sprays, but supplies no labour or machinery. In the case of contractors, the farmer pays the cost of the seed, fertiliser and sprays, plus the contract price of the tractor services (which includes the labour to operate it).

Sometimes, contractors do not arrive on time* – we have allowed 10% off the expected gross margin for lack of timeliness when contractors are used. Also, we have added $8/h for contractors' charges. In Tables 13.4 and 13.5 we show the relevant factors which underlie the decision. Note: The hourly costs are those derived in Table 13.2.)

In this first case, ownership is more profitable than both contracting and share farming, mainly because the tractor is getting plenty of use (500 h/year).

Let us now examine what may happen on a farm when there are 7.5 ha, following the same rotation. Here tractor use will be only one quarter of that in the above example, i.e. 125 h. In this case, tractor costs rise to $21.30/h (see Tables 13.3 and 13.5), whilst contractor costs remain the same as for the 500 h job, i.e. $19.1/h.

In this case, it is slightly better to use contractors than to own a tractor, because of the low number of hours of use and high overhead costs per hour. Share farming, if the proceeds and costs are shared as we have assumed, is better than either owning or contracting.

One feature of owning a tractor (or any other machine) is the possibility of contracting to other farmers after work on the home farm has been

*Let us say that the hourly cost of running a tractor is greater than the contract price per hour, but the farmer prefers to use ownership rather than contract services because (to him) it is safer. The contractor may not arrive on time. Why not pay the contractor a special incentive bonus to make sure that he does arrive on time and so avoid yield loss? Even with the extra cost of the bonus, it may then still be more profitable to use contracting services.

Table 13.5. *Cost of machine services for 125 h work*

	Own ($)	Contract ($)	Share ($)
Gross income (less harvesting costs)	7500	6750	2500
Costs			
Variable growing costs, excluding tractors (seed, fertilisers, etc.)	3000	3000	1000
Tractor costs for cultivating (100 h)	2130 (100 × 21.30)	1910 (100 × 19.1)	
Tractor costs for transport (25 h)	532.50 (25 × 21.30)	477 (25 × 19.1)	
Tractor driver	1000	Nil	
Sub-total of costs	6662.50	5387.00	
Surplus to pay other costs and profit	837.50	1363	1500

completed. This is a common system in many developing countries. Small farmers cannot afford the capital investment in the machine, so they hire contract services for ploughing, sowing, harvesting and threshing from other, mechanically minded farmers, who own medium-sized farms. Sometimes government and semi-government bodies run machinery-hire and contracting services, which can be disastrous if the machines are not maintained properly, if they do not arrive on time, and if (because many people are competing for the services) bribery of officials becomes the normal way of achieving timeliness of operation (this would probably be an overhead cost.)

Using draught animals

In the case of buffalo, or ox (bullock) power, it is worth noting that there are many similarities with the example just shown, which was based on the tractor. Some assumptions we have made are:

two oxen are needed (a young ox – or buffalo – trained for work, costs $1000);

they have an effective working life of 5 years, when they are sold for $600 (the replacement oxen will cost $1200 each);

the average value of an ox is $(1000 + 600)/2 = \$800$;

the amount to charge (in today's dollars) for depreciation so that there will be enough money set aside each year to replace the ox will be:

expected replacement cost $1200;

expected sale price $600;

total depreciation and replacement = $600.

If investing a sum of money each year at 2% real interest to grow to the $600 needed to replace an ox in 5 years' time, then the sum to set aside each year is

$600 × 0.1921 (sinking fund factor) = $115.

In other words, $115 invested at interest each year plus the interest reinvested, will grow to $600 in 5 years' time, when the ox has to be replaced. So the annual depreciation and replacement cost of the ox is $115. Other assumptions are:

	Cost ($)
Insurance/head	50
Interest rate 2% of $800/head	16
Feed costs/head	200
Veterinary costs/head	20
Repairs to equipment (harness)	10

The annual costs of owning and using two oxen or buffalo are shown in Table 13.6.

As well as overhead and variable costs, there will probably be hidden costs in timeliness of operations and standard of work, which would add to the real cost of oxen (or buffalo) services.

As we noted earlier, we are primarily concerned with identifying the *factors* which are important in making decisions about getting machine services. It is up to the individual and/or his adviser to attach the relevant numbers (i.e. costs, yields) to the budget framework and headings we present here. For example, some farmers will charge for their own time, others won't. Also, there will be differences in the allowances made for untimeliness and quality of work done by both contractors and draught animals. And, while some farmers will make an interest charge for the money they have tied up in machinery and draught animals when comparing alternative sources of machinery service, others will not regard this as important.

Table 13.6. *Annual cost of using 2 oxen (bullocks) or buffalo*

	Cost ($)
Overhead costs	
Depreciation and replacement (115×2)	230
Insurance	100
Interest on average value ($800 \times 0.02 \times 2$)	32
Total overhead costs	362
Variable costs	
Feed	400
Veterinary and other	60
Total variable costs	460
Total costs/year	822
Total operating costs/year	
Hectares per year worked	5
Cost/ha	164[a]

[a] In the tractor example, using contractors at $19.1/h, at 10 h/ha, the cost/ha was $191.

Other factors to consider when making decisions about machinery services

Timeliness
Many farm operations must be performed within a limited time period if excessive losses are to be avoided. This applies to most fallowing, weeding, sowing and harvesting operations. A contractor who takes on too much work for the capacity (amount of work which can be completed during the season) of his machine, finds that his period of operation is extending outside the optimum time, thus incurring the penalty of lowered quality and quantity of product. During the harvesting of grain, or hay, this loss can be substantial. As it is income forgone (i.e. it is a penalty cost), it must be charged as a cost against the contract service.

To balance this, some of the available contract and share farm services provide expert 'know how' through specialisation in their particular type of work. Providers of contract and share farm services are often able to provide a better quantity and quality of product than the farmer does. Expert 'know how' can reduce the total cost per unit of product which results.

Opportunity cost of capital
The type of service chosen will also depend on the availability of capital and the rate of interest which the owner decides to charge on it. The interest rate used should be the one which the capital could be earning were it invested in the best alternative (such comparisons are usually in real terms); on some farms this can be as high as 8–10%. A high 'opportunity' rate of interest usually means that services other than ownership are best.

Machinery syndicates
The main advantages of machinery syndicates (group ownership) are that they reduce the overhead cost of machine ownership, frequently allow profitable use of limited capital funds, use otherwise idle machine capacity, and make available machine services which would otherwise not be available.

Disadvantages are disputes and arguments, or some members being disadvantaged in relation to the maintenance work, the allocation of repair costs, the order of machinery use; also, uncertainty as to whether work of the desired quality will be carried out (and on time).

Another way of increasing the use of machinery, which is gaining in popularity, is where the machine remains the private property of one individual who hires out the idle capacity for cash. The fee is sometimes on a non-cash basis, with the user being obliged to return an equivalent service with some other service.

The main message from our discussion so far is that it matters less which form of machinery service is selected, than that each service does its basic job (growing a good crop) efficiently. In general, the machines whose services are most economic to obtain by methods other than ownership are those which:

have a low annual use on the farm;

require a large investment (e.g. earth-moving equipment);

perform operations in which timeliness is *not* crucial.

When the decision about ownership is not clear-cut, the following points have to be clarified:

the annual use to be made of the machine;

likely changes in costs of contract versus ownership (e.g. contract prices may decline relative to ownership costs as more contractors compete for work in an area);

the value of timeliness of the operation;

the value of the quality of the completed service;

the rate at which profits can be earned on alternative investments;

the approximate opportunity cost of one's own labour force at the time the machine would be operated by that labour;

(possibly) the rate at which income tax is being paid.

What sized machines?

What sized machines should a farmer buy? This decision usually depends on a combination of farmer's needs, personal preferences, economic position and sales pressure. The logical approach to deciding how large a machine should be is to ensure that it is large enough to get all the most important field operations done well, on time, and that it has enough work to keep the overhead costs per hectare down to a 'reasonable' level.

If the yearly work programme is to be completed, a tractor of minimum capacity for the farm would theoretically use all the available days to complete the programme. In practice, some 'slack' to handle emergency work, or work unaccounted for, or extra working or resowings, has to be allowed for in budgeting annual hours of use.

The value of timeliness of operations should be the first consideration when choosing the size of machinery, as loss of income from untimeliness alone can more than cover the extra cost of a high-capacity machine in the first year. This particularly applies to the weed control, sowing and harvesting operations. By comparing the relationship between (i) the cost of untimeliness and (ii) the added overhead costs of larger capacity machines, a definite minimum capacity can be determined. The other factors, already discussed, will determine optimum size. There is not much information available on the value of timeliness of operations; some recent studies in Africa indicate yield gains from timely sowing of groundnuts and sorghum to be as high as 50% and for timely weeding to be 50% for groundnuts and over 100% for millet. Australian researchers have found that untimeliness at fallowing can cause losses of 200 kg/ha of grain, while untimeliness at harvesting can cost 400 kg/ha. The major

operations with high 'penalty' costs for untimeliness are:

time of fallowing (delays mean less moisture stored, hence lower crop yields);

speed of working fallows (weeds have less chance to transpire moisture if killed quickly);

time of sowing (delays lead to lower yields from either weed competition or loss of moisture);

harvesting (rain damage, shaking and hail loss).

Farmers recognise the cost of untimeliness and are prepared to pay a price to minimise it. Thus the oft-heard claim that farmers are 'over-capitalised' in machinery needs to be viewed in terms of timeliness. The economics of decisions on machine size depends mainly on:

the cost associated with failure to perform the operation on time (i.e. the value of timeliness of operation);

the value of labour saved (if any) by the larger machine;

the difference in the annual overhead cost of the large and small machines;

the number of hours which different tractor sizes would take to complete the cropping programme;

the availability of, and alternative uses for, capital;

the opportunities for contracting.

The conditions which favour the use of a large machine are: large risk from weather or untimeliness; high labour costs; high annual hours of use and low opportunity interest rates on capital.

The elements of the question: 'what sized machine or collection of machines is best' can be shown on a diagram. The key factors involved are the number of hectares, the expected costs of owning and operating the different sized machinery system, and the costs due to untimeliness of operations which are expected to occur with different sized machinery (see Fig. 13.4).

Questions

1 How important is timeliness of cropping operations in your area?

2 Do farmers you know get their machinery services

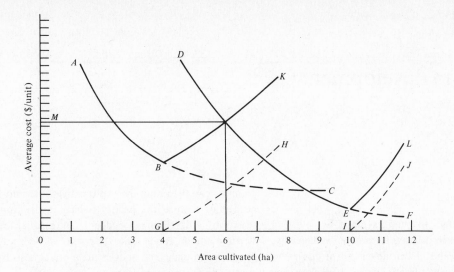

Fig. 13.4. Total costs of small versus large sized machinery.
Line *ABC* represents the average costs per hectare of
owning and operating the small machinery. Line *DEF*
represents the average costs per hectare of owning and
operating the large machinery. Lines *GH* and *IJ* represent
the timeliness costs associated with the small and large
machinery, respectively. Lines *ABK* and *DEL* are the total
cost curves for the small and large machinery, respectively.
Please note the following points.

(i) Beyond 4 ha, with 'small' machinery, there are costs
which are due to untimeliness of operations. This raises
the total costs of the small machinery system.

(ii) Beyond 10 ha, with the 'large' machinery system, there
are costs which are due to untimeliness of operations,
raising total costs.

(iii) Beyond 6 ha, the larger machinery system costs less/ha
than the smaller one. At 6 ha, they break-even at cost
M.

(iv) For less than 6 ha the smaller system is cheaper.

by ownership, contracting, share-owning or by
mutual exchange?

3 Do some sums with a local farmer's figures to
compare the annual costs of obtaining some machinery service from a couple of different sources, in
a couple of different ways. Why must these sums be
done in today's (real) dollars, using real interest
rates?

4 Once you have calculated which form of machinery
service is the cheapest, if ownership is involved,

work out how much the farmer has to set aside each
year in order to replace the machine. Why must you
now use nominal dollars and nominal interest rates?

5 Work out whether contracting is a viable proposition for local farmers, at the current level of contract
services, prices, quality of job and timeliness of
operations.

6 Work out whether there is scope for one of the better
local farmers to purchase a machine and provide
contract services to the district.

14

Farm development

Introduction

Development refers to changes made by investing capital in a farm to increase its productive capacity, profitability or value. Farm development can mean many things. It can be a small change taking place over a short time and financed from annual profits. More commonly, it is carried out over at least 5–6 years and is financed by relatively substantial borrowings for capital investment.

Why develop a farm? This is usually done in order to maintain or increase income, and/or to increase the value of the farm assets. The benefits of development are reaped by selling increased crop production, converting extra feed into saleable livestock and products, selling assets at a higher price, using improved borrowing power, or by a combination of these. Some examples of farm development are:

clearing of bush, fencing and cropping or pasture improvement on an unimproved area of the farm;

developing an area for irrigation;

establishing an intensive farming sideline.

Development budgeting involves a series of annual cash flow budgets, generally drawn up until the development is complete and the annual results are stable, i.e. have reached a 'steady state'. Development budgets are designed to show expected future costs and returns associated with a development programme. They are the basis for deciding whether or not to undertake a project, for obtaining finance, and for monitoring it once it gets under way.

There are usually many alternative ways of carrying out a proposed development project. Each has different risks and costs, and some forethought and pre-

planning can help rank these alternatives. If credit is needed to finance the project (which is the usual situation), a development budget is an essential first step in obtaining funds. Once the project is under way, development plans and budgets are a useful basis for control and revision of plans.

Development budgeting uses the techniques of cash flow budgeting. Fairly detailed physical planning of a development project is first necessary to be sure that it is technically feasible before any budgets are drawn up.

Return on extra (marginal) capital

Where a programme of property development is considered, the probable return on investment is an important preliminary criterion for decision-making. The technique is discussed in detail in Chapter 10, but we will review the main points here. To assess returns on extra (marginal) capital invested:

(i) add up the total capital costs of the development (including extra livestock and plant);

(ii) estimate the extra 'profit' (after extra interest plus tax) which it is expected will result once the improvement has stabilised at the higher level (assumptions have to be made about yields, prices and costs; the partial budget is the technique to use in calculating extra 'profit' – (see Chapter 10);

(iii) express (ii) as a percentage of (i) to obtain the return on extra invested funds.

Some points to consider are:

If the project appears satisfactory on this assessment, (more than 10% return on marginal capital), a detailed year-by-year cash flow budget is needed. Where borrowed funds are to be repaid over a long period the effect of inflation on costs, and hence on net cash flow is important.

If development is carried out with otherwise surplus or idle machine and labour capacity, then only the variable machine and cash labour costs and the extra use depreciation of the machines are charged, not the capital cost of the machine.

Steps in development budgeting

The first step (*A*) in budgeting a development proposal is to collect physical and financial details. Useful physical information includes a farm map, stock numbers, soil types and nutrient status, fertiliser history, crop and animal yields, plant, available labour and skills, and the state of structures. Financial information which is useful includes past operating costs, assets and liabilities, debt servicing obligations, prices received, cash available for development, and borrowing capacity.

The second step (*B*) in development budgeting is to plan the physical development in detail. It is important to carefully check that the plan is physically achievable, the sequence of events logical, the progressive production increases not overestimated and that the plan can be financed.

A field development plan covering clearing, cropping, water and building developments is vital when estimating annual financial commitments. A field cropping or rotation plan is applicable where the development involves the planning of rotations. This could be used in an irrigation and cropping-land development, or in assessing the changeover from a predominantly livestock enterprise to cropping. A schedule of plant and improvements on hand, and additional plant and improvements required for the development programme must be included in the development budget. A labour plan indicating the extra labour requirements in terms of skills and numbers due to development is also needed. A thorough analysis of labour requirements for any development is most important in determining profitability.

The third step (*C*) is to do the budgets. Dollar values are placed on the physical plan to estimate capital requirements, cash income and costs. The income from a development is generally derived from increased crop production and livestock build-up. The costs of the development need to be estimated and include all relevant operating, capital, interest and finance costs.

The particular form of cash flow budget which is done depends on its purpose. To compare two projects, cash flow budgets with 'real' dollars are made and discounted to net present value. If a project has already been chosen, income and cost flows are calculated on a yearly basis, in nominal dollars up to the year in which they become 'stable'. That is, up to when the development plan has been largely achieved and has settled into a regular pattern of production, income and costs. This is known as the steady state. The nominal (not constant) income and cost figures are then used to construct a cash flow budget for the period up to the steady state.

Often it is only the costs and returns from one part of a farm which are analysed – the cash flow constructed is only for that particular project. Such cash flows are useful for analysing particular projects but do not give the whole farm picture. From the annual and cumulative net cash flow, estimates of loan requirements, peak debt and the overall soundness of the project can be made.

Major difficulties with development project budgeting are in deciding on the life of the project, and what 'salvage' value to use. With some projects, say orchards, the life of the stand of trees may be well known. Other projects, like land development may have a seemingly infinite life in its improved state, if properly managed. In these cases one useful approach is to select a planning period of some finite length and estimate the end value of the project at the end of that period (say 10 years). In doing so the aim is to estimate what the project might be worth in 10 years' time. The salvage, or end, value is often very much a guess, depending on whether the improvements have a known rate of depreciation and on the vagaries of future markets for productive assets and commodities.

The 'net cash flow' shows how much new borrowing (if any) had to be made each year. With most projects, this is likely to be relatively large in the early years. In later years no new borrowing is usually needed.

The difference between cash spent in any year (initial capital input, new capital investment, minimum acceptable funds for living requirements, tax, variable costs, interest, principal payments, plant replacements, but not depreciation) and cash received (e.g. sales of animals, produce, capital items such as land or plant, but not inventory increase) is termed the annual net cash flow. The extra market value of land, plant and stock on hand at the end of the planning period, is known as the salvage (or terminal) value of the project. For the appraisal of an investment, the salvage (or terminal) value can be regarded as cash received, and money owed at the end of the project is regarded as cash spent.

The best way to demonstrate the three parts of development budgeting – *A*: physical and financial details, *B*: the physical plan and *C*: the budgets – is to work through an example.

A: Physical and financial details

A farmer owns 6 ha, 4 ha of which can be cropped. The remaining 2 ha are not cleared and the farmer is thinking about clearing the land and using it for growing crops. At the moment, the farm is breaking even and there are no debts, so that all cost increases will be due to the development programme.

The development involves clearing the trees from the land, removing tree roots, cultivating and growing a first, ground-breaking crop, clearing regrowth, fencing and forming the land into irrigation beds for cropping. Cropping has been chosen in preference to livestock activities because it offers higher returns. It is planned to rotate crops of irrigated corn with a grain legume crop, chick peas. This rotation gives the highest total gross margin from all the alternative crop rotations which were technically feasible. Total gross margin from the rotation is expected to be $2000. There will also be some extra overheads ($300). Total capital required is $3000. The first corn crop will be planted in Year 3 of the project.

Estimation of return on extra (marginal) capital
We have shown that the rate of return on extra capital is a useful preliminary screening device for appraising investments. Let us now calculate the return for funds invested in this project.

Calculation of return on extra (marginal) capital

Extra capital needed (over 3 years) $3000

Change in 'profit' after interest in Year 3 compared with the present situation:

	$	$
Extra cropping income		
Corn	2300	
Chick peas	400	2700
Extra cropping costs		
Overhead costs	300	
Variable costs	700	1000
Extra interest (3000 × 0.10)		300
Change in profit (before tax)		1400
Tax (say 25¢ in $1)		350
Net profit		1050
Rate of return on marginal (extra) capital		$\frac{1050}{3000} \times \frac{100}{1} = 35\%$

This rate of return on extra capital suggests that a detailed cash flow is worth calculating.

B: Physical plan

In order to do a detailed cash flow budget, the physical aspects have to be carefully planned (Table 14.1).

C: The budgets

Capital requirements

Year 1	Year 2	Year 3	Total
1000	1400	600	3000

Salvage value of the project at the end of Year 8
When evaluating a development project it is essential to consider the change in asset value, as well as the net cash flow. In this example, we assume that each $1 spent on improvements has a salvage value of $0.80.

Development budgeting, and taking account of 'time is money'
Appraisal of development projects becomes more complicated when the value of money over time has to be taken into account. This is shown in Table 14.2. The 'complicating' factor is that the projected cash flows

Table 14.1. *Physical plan for a 3 year development programme*

Time	Action	
Year 1	Clearing	
Year 2	Clearing	
	Fencing	
	Cultivating	
	Planting	
	Harvesting	
	Cultivating	
Year 3	Irrigation beds formed	
Year 3 and after	Steady state reached	
	Cropping corn and chick peas	
Yield after year 3	Corn	3 tonnes/ha
	Chick peas	0.4 tonnes/ha

Table 14.2. *Cash flow development budget (real, constant dollars – no interest charged in cash flows)*

		Years							
		1	2	3	4	5	6	7	8
A: Cash income before loans									
Corn		0	0	2300	2300	2300	2300	2300	2300
Chick peas		0	0	400	400	400	400	400	400
Salvage value									2400
Total receipts		0	0	2700	2700	2700	2700	2700	5100
B: Cash paid									
Extra variable costs		0	0	500	500	500	500	500	500
Extra overhead costs		0	0	200	200	200	200	200	200
Capital Development Costs									
Clearing		1000	500						
Fencing			600						
Net costs of first-up crop			300						
Irrigation					600				
Total costs		1000	1400	1300	700	700	700	700	700
Before-tax annual cash deficit/surplus		− 1000	− 1400	+ 1400	+ 2000	+ 2000	+ 2000	+ 2000	+ 4400
Plus tax (25¢ in $1)		0	0	− 300	− 480	− 500	− 500	− 500	− 500[a]
C: Annual net cash flow (A − B)		− 1100	− 1400	+ 1100	+ 1520	+ 1500	+ 1500	+ 1500	+ 3900
Net present value	*Total*								
Discount rate 0%	+ 8620	− 1000	− 1400	+ 1100	+ 1520	+ 1500	+ 1500	+ 1500	+ 3900
Discount rate 15%	+ 2897	− 870	− 1058	+ 723	+ 869	+ 746	+ 648	+ 564	+ 1275
Discount rate 45%	+ 55	− 690	− 666	+ 361	+ 344	+ 234	+ 161	+ 111	+ 200

[a] Tax not charged on salvage value.

Table 14.3. *Cash flow budget (nominal dollars)*

	Year							
	1	2	3	4	5	6	7	8
A: Cash income before loans								
Corn	0	0	2300	2400	2500	2600	2700	2800
Chick peas	0	0	400	450	500	550	600	650
Salvage value								2400
Total receipts	0	0	2700	2850	3000	3150	3300	5850
B: Cash paid								
Extra variable costs	0	0	500	600	700	800	900	1000
Extra overhead costs	0	0	200	300	400	500	600	700
Capital development costs								
Clearing	1000	500						
Fencing		600						
Net costs of first-up crop		300						
Irrigation			600					
Finance costs								
Interest on cumulative cash deficit @ 10%	0	110	276	191	61	0	0	0
Total costs	1000	1510	1576	1091	1161	1300	1500	1700
Before-tax annual cash deficit/surplus	− 1000	− 1510	+ 1124	+ 1759	+ 1839	+ 1850	+ 1800	+ 4150
Interest on annual cash deficit (10%)	+ 100	+ 151	0	0	0	0	0	0
Tax (25¢ in $)	0	0	− 281	− 452	− 486	− 500	− 500	− 500[a]
C: Annual net cash flow (A − B)	− 1100	− 1661	+ 843	+ 1307	+ 1353	+ 1350	+ 1300	+ 3650
Cumulative bank balance	− 1100	− 2761	− 1918	− 611	+ 742	+ 2092	+ 3392	+ 7042

[a] No tax on salvage value.

from the farm development need to be discounted back to present values to find out what the proposal is really worth in today's dollars. Detailed project appraisals, which involve discounting future income streams back to present values, require the following steps.

First, cash flow budgets are prepared. These are based on alternative physical plans for the development being contemplated. Values for yields, costs and prices are assigned on the basis of forecasts and market outlook information. Today's money values rather than future nominal, inflated money values, are used. The budgets are projected far enough into the future to reach relatively stable cash flows, e.g. 6–7 years.

The next step is to discount these expected future cash flows back to present values. This is done by discounting at the chosen interest rate, (e.g. 15%), and then totalling, thus giving the total present value of the project.

In our example (Table 14.2) the project has a positive net present value at 15% discount rate. It will not become negative until a discount rate of over 45% is used. It thus should be quite a 'profitable' project. A cash flow development budget now needs to be done using nominal dollars, to find out the timing and magnitude of cash deficits and surpluses.

Cash flow development budget

In this example, (Table 14.3), income and costs are in nominal (inflated) dollars. On the 'cash in' side, gross margins and salvage values are included and 'cash out' contains development costs, tax and extra overheads. The resultant net cash flow figure is simply the difference between 'cash in' and 'cash out'. The cumulative bank balance indicates the extent of total borrowings which will be necessary. Interest is charged on both the cumulative and annual cash deficit. The figures in Year 1 represent cash in and out during Year 1 and are the total on the last day of the year.

It can be seen from the cumulative bank balance figures, that it takes 5 years before the bank balance comes into credit. Another notable feature of the budget is that when account is taken in Year 8 of the salvage value of the extra investment in improvements, there is quite a marked jump in the credit balance. Provision has also been made for inflation by increasing costs of inputs, with a (slower) rise in product prices as well.

Questions

1 What are the three major steps in development budgeting?

2 When do you use real values and when do you use nominal values in development budgets?

3 Draw up a development budget for a project on one of the local farms (in today's dollars). Will it work? Have you allowed sufficiently for risk? Does it pass all the tests of feasibility?

4 Now, draw up a cash flow budget for the project using nominal dollars. What is the peak debt? When does it start to bring in more than it costs? How much would the farmer need to borrow? Can the project be financed?

5 What do you understand by the term 'salvage value'?

15

Farm credit and finance

Theory

General

Getting and using agricultural credit is an important part of money management, and good money management can help the farm business to grow. Borrowed money can be used to increase both farm production and wealth. However, lenders are sometimes a bit sceptical about farmers who borrow from them. As one experienced farm lender told the authors, 'Many receivers find it hard to repay, thinking it's pocket draining, forgetting the getting'. There are two broad groups of farmers who need credit:

(i) those who have started developing their farms without credit but find that savings are too low or too slow to let them carry on the development;

(ii) those whose production is so low that they cannot save any money to increase production, and despite many opportunities and keenness to progress, cannot increase output without getting a loan from somebody.

To whom do farmers turn to obtain credit? Possibilities include: family, friends, merchants and banks. Alternatively, they can join a cooperative (which is partly government-backed); sometimes, in such ventures, the loans are not wholly repaid.

Lenders of money usually expect some payment or reward. The cost of borrowing or hiring money is called interest. In countries where it is against religious custom to charge interest, a borrower may still have to reward the lender in some other way. Human ingenuity has led to the creation of numerous devices to handle the ethical problems involved.

The lender expects to be repaid in the future. When capital is lent, there is always a risk that it will not be repaid. The lender generally wants some reassurance against the loss of his capital. Such reassurance may be simply trusting the borrower's promise to pay, or something more tangible. Often, the lender will get the right to take ownership of a borrower's asset if he does not repay the loan. The term 'mortgage' refers to this situation. The borrower 'mortgages' his land, stock or any other asset of value to the lender. If the borrower defaults on the loan, the lender takes possession of the asset and may sell it to recoup his money.

In situations where the borrower has very little to give the lender as security in the form of assets, then the lender has to rely on the productive capacity of the land. The merchant–trader will lend money to the producer as long as he gets first claim on the crop. The amount of the crop the merchant–lender takes adds up to the value of the money lent plus the interest on the loan. The size of the interest charge depends upon how desperately the borrower needs the funds, the risk the lender takes in lending the money, and the expected rate of inflation – that is why the interest charged is sometimes 30–40%.

Borrowed funds should be used productively so that the debts can be repaid when they are due. Sometimes credit is used to smooth out uneven household money flows. For example, borrowing to pay for items such as clothes, books, taxes, weddings, and repaying the debt after selling some of the crop. The repayment of borrowings needs to be closely related to the expected pattern of the rise in farm output and income which will (hopefully) result from the use of the funds. This is all very well in theory, but what happens in situations where:

there is a very low level of saving in the country, hence low levels of funds for credit and high interest rates;

there are few established financial institutions;

farmers have little or no collateral (i.e. assets which can be borrowed against);

there is high risk and uncertainty about the outcome of the use of borrowed funds;

there is a high risk of default on the loan;

there are high costs of administering loans, relative to the profit that can be made from lending (this is often the case with loans to small farmers);

skilled lending personnel are rare and have little knowledge of farm financing;

farm incomes are small;

farm areas are fragmented, vary greatly in quality and area and may not even be known accurately;

land tenure may be based on social custom, communal or tribal ownership?

The issues listed above are very real ones in lending in many tropical countries. That is why we have partly aimed our book at rural lenders: to help them to understand the economics of farm management at the farm level.

Equity and debt servicing

In answer to the questions of to whom to lend, how much, and how much to borrow, there are some general principles of money management. If a farmer has some money, and owns some land and equipment which he can use to grow a crop or which he would sell for money, then he has some *assets*. For example, suppose a farmer has –

	Value ($)
Cash and savings	30
Stock he could sell for	150
Equipment he could sell for	20
Total assets	200

Another name for assets is 'own capital'. If he borrows $50 and buys an asset which can later be sold; e.g. a plough (but not a wife), he now has a total capital of

$250. Of this $250 total capital his share, called equity, is $200. His equity ratio is:

$200/250 = 0.8$.

Equity, which is also called 'net worth', is the net amount which would be left if the assets were sold and all debts paid. It is usually expressed as a percentage of total assets, thus:

$$\text{Equity } (\%) = \frac{\text{Total assets} - \text{Total liabilities}}{\text{Total assets}} \times \frac{100}{1}.$$

Calculating equity in some cases is difficult. The main problem stems from values placed on various assets such as land and improvements, e.g. cleared land, a building, a watering system. It is difficult to measure the contribution made to total equity by any physical development. Thus, $1 invested in land clearing may not have a salvage or sale value of $1. This fact explains why equity sometimes declines during a development programme. Most other assets, such as livestock and machinery, are readily valued. Generally, the more fixed an asset, the more difficult it is to value. Equity can be eroded by a serious drought or a major decline in price.

The higher the equity percentage, the easier it is to 'service' borrowings. The lower the annual cash surplus before interest and the loan repayments (which are called principal) the higher the minimum safe equity level. This ability to service a debt depends on the level of the annual cash surplus before deducting interest and principal payments which result from the year's operations. It is calculated as follows:

A. Gross cash income
   Less
B. Cash living expenses
   Net plant replacement costs
   New capital investment from
     current income
   Taxes
   Total cash outgoings
C. $(A - B)$ Annual cash surplus (net cash
     flow) (available for interest and
     principal payment)

Note that in calculating the annual cash surplus, non-cash items like depreciation, operator's labour, and others, such as the rental value on the house, are omitted. The higher the annual cash surplus, the higher the debt that can be serviced and the lower the critical equity level needed to be safe from going bankrupt.

In most farm development schemes there will be no positive net cash flow until the 4th or 5th year. In making budget projections about the expected level of annual cash surpluses, it is necessary to provide for the possible effects of drought, price declines and drastic cost jumps. It is important not to allow the equity to drop to dangerous levels, and to have contingency provisions in the loan repayment arrangements.

The following table illustrates how cash surplus minus interest, rather than equity alone, determines the ability to service a loan.

	Case *A* (low equity)	Case *B* (high equity)
Cash surplus, minus interest	200	40
Loan	2000	2000
Time to pay (years)	10	40

Though *A* has low, and *B* high, equity, *A* is a much more attractive proposition than *B*, even though from the security point of view, *B* is better than *A*. Security alone is thus a dubious basis for advancing funds.

Growth profits and leverage

A key aspect of credit use is the role which credit can play in the growth of a farmer's wealth. To do this, credit must increase earnings from the farmer's resources. The increase in earnings from the use of credit must exceed the full costs of borrowing, and allow the loan and interest to be repaid. The relationship between debt and equity, and rate of return on borrowings, help determine what the farmer will eventually be worth. First, let us look at 'profit'. As we indicated earlier, the term 'profit' can have many meanings. One definition is:

Profit = interest earned on assets minus interest rate paid on debts.

This can be written $P = (rA - iD)$,

where r = interest rate earned on assets, A = assets, i = interest rate paid on debts, D = debts.

Next, taxes and spending on personal consumption out of profits have to be deducted. Consumption and all other off-farm spending can be called the consumption rate. This is subtracted from after-tax profit. Whatever is left is an increase in savings or equity. This can be called the 'growth profit' and represents the growth which has occurred in savings or equity (net worth). Thus:

Total return on assets
Minus
(i)   Interest on debt
(ii) Tax
(iii) Personal consumption
Leaves
Savings, or growth profit

One concern is how the borrowed money can be used to earn a rate of return greater than the cost of borrowing. Whilst the rate of return in farming can fluctuate a great deal, interest charges have to be paid regardless of whether the rate of return is high or low.

The principle of increasing risk tells us that when there is an increase in debt to equity ratio (which is also known as leverage, or gearing) then the risk of going bankrupt increases. Leverage has this effect: when things go well and rate of return exceeds rate of interest paid, growth occurs faster than would have been the case with no debts and no leverage. High leverage also means that, when things go badly and rates of return do not exceed interest costs, equity is eroded at a far quicker rate than the rate at which it would have grown had returns been good.

As well as being affected by variable rates of return, growth is affected by unexpected changes in debts and assets. A change in asset values can be caused by changes in market conditions, or a run of above- or below-average seasons. If asset values fall the leverage automatically increases; for example:

Today		*6 months later*	
Assets	500	Assets	300
Debts (*D*)	200	Debts (*D*)	200
Equity (*E*)	300	Equity (*E*)	100
Leverage: $\dfrac{D}{E} = 0.66$		$\dfrac{D}{E} = 2$	

This change in leverage could mean the difference between farm business survival and ruin.

So, growth of the farm finances is affected by:

rate of return on total resources;

interest on total debt;

debt to equity ratio (or leverage);

rate of personal consumption of profit;

rate of tax.

Inflation and interest

In countries where there is inflation (how many have none?) then the interest rate charged on funds is made up of two parts: (i) one part to cover the effects of inflation, (ii) one part which is the gain or reward to the lender. This is the price the borrower pays for getting the use of scarce funds, plus an allowance for the risk that the lender takes that he might not be repaid. It is called the real rate of interest. When someone with some funds makes a decision today to use that money in some way for one year, his thinking would go something like this:

> I have $100. One way I could use the $100 would, I expect, bring me in $130 at the end of 1 year. I will have $30 more then, but the currency will probably inflate at 20% over the year, and if it does, then my original $100 plus the $30 I have gained will each be worth 20% less when I get the money in 1 year's time! With 20% inflation, $1 in 1 year's time will be worth 100/120 or 0.833 of $1 today. My real gain will be:
>
> Today's $100 = $100 purchasing power
> 1 year's time $130 = $130 × 0.833 purchasing power = $108.30
> Real rate of return = $8.30 or 8.3%.

That is, the real rate of return on the $100 will be 8.3%, whilst the nominal rate of return is 30%. Inflation accounts for the difference. Lenders are interested in their original funds growing in buying power (real terms). Their investment decisions depend upon what they expect to happen.

In this example, if things turn out as the lender expects, then this lending option will bring a real return of 8.3%. The lender might think:

> I could get a real return of more than 8.3% by lending my $100 in the money market. If inflation

is 20% this year and I charge an interest rate of 35% (giving a total return of $135), my real return will be $135 × 100/120 = $112.50, or 12.5%.

So, he lends his $100 at 35% interest. This interest is made up of 12.5% real return (the price of money which is due to the scarcity of funds and risk) and about 20% to cover the effects of the expected inflation on both the original $100 and the earnings. The market rate of interest (m) is approximately equal to the real rate (r) plus the inflation rate (f):

$$r + f = m,$$
$$12.5\% + 20\% = 32.5\%.$$

More precisely,

$$m = r + f + rf,$$
$$0.125 + 0.2 + (0.125)(0.2) = 0.35.$$

In practice, the inflation rate (f) is not known before an investment and so it is the expected inflation rate which helps form market rates of interest.

Loan repayments

Most lenders need repayments of principal (the sum borrowed) as well as interest payments. The two main types of loan repayments are term loans and amortised loans. Term loans are usually either for farm development or farm purchase. This type of loan is generally available for from 5 to 10, possibly 15 years. With a term loan, principal is repaid in equal annual or half yearly instalments. Simple interest is paid separately on the outstanding principal and this reduces annually as the principal is repaid.

The ability to meet all debt-servicing commitments as they occur is critical when considering how suitable a particular loan is. With a term loan, the vulnerable period is during the first few years when repayments are greatest. With amortised loans, repayments are made in equal annual instalments consisting of both principal and simple interest. As principal is repaid so the interest content of each payment is reduced, allowing more capital to be repaid. These loans are usually more expensive than term loans because the average amount of principal outstanding is higher.

A strategy sometimes employed to improve the chances of the successful use and full repayment of a loan is a principal repayment holiday at the start of the loan. This is very convenient if the initial costs of an investment are high in the early years and the cash flows do not become positive until quite some time later. The effect of a principal repayment holiday is that the initial debt-servicing obligations are greatly reduced. The total cost of the loan is usually greater (in nominal terms at least).

Flat interest and simple interest

With hire-purchase (HP) and lease agreements, a different method of lending is used. The interest is charged on a 'flat' basis, i.e. interest is paid on the amount originally borrowed, and not on the balance that is owing, regardless of the fact that part of the loan is being repaid each year. For example, $100 for 4 years at 15% *flat*, incurs the following annual repayments.

	Year 1	Year 2	Year 3	Year 4
Interest	15	15	15	15
Principal	25	25	25	25

Total interest paid = $60.
With simple interest the interest charge is calculated by multiplying the average amount of principal outstanding during the term of the loan by the interest rate. If simple interest were charged, the repayments would be:

	Year 1	Year 2	Year 3	Year 4
Interest	15	11.25	7.5	3.75
Principal	25	25	25	25
Debt at end	75	50	25	0

In this case the interest paid is $37.50, which is $22.50 less than when flat interest is charged. To convert the flat interest rate that the hire purchase company quotes, to its effective simple interest equivalent, use the formula:

$$i = (2ft)/(t+1),$$

where i = simple interest rate equivalent as a decimal, f = flat rate of interest, t = number of equal repayments

or instalments, 1 = the number 'one'. Using our example:

f = 15% (0.15)
t = 4 (payments, one per year)
$t + 1 = 5$.

Then,

$i = [2(0.15 \times 4)]/5,$
$= 1.2/5 = 0.24$ or 24%.

For any flat rate of interest on a loan of 3–4 years' duration, the rule of thumb for conversion to a simple interest rate is that the equivalent simple interest rate is about two-thirds more than the quoted flat rate. When other compulsory charges such as tax, insurance and sometimes 'front end load' are added to the HP agreement, then the effective simple rate is nearly twice the flat rate ('front end load' is a charge sometimes made by the lender just to set up the particular finance deal).

So, in summary: with a term loan it is the high debt-servicing required in the first few years that is critical in assessing the ability of a farm to service the loan. An amortised loan, by equalising the annual payments, is more easily serviced than a similar term loan. Principal repayment holidays can be a big help, and HP loans, using 'flat' interest, are expensive.

What steps can be taken to reduce repayments to an acceptable level? This is a question most borrowers ponder. Generally, the length of the loan has more effect on the level of annual repayments than the interest rate. There is more scope for reducing the size of annual repayments by increasing the time span on a loan, rather than by reducing the interest rate.

In Table 15.1 below, the annual repayment on a

Table 15.1. *Annual repayments under different loan terms*

Length of $200 loan (years)	Interest rate	
	10%	11%
7	41.08	42.44
10	32.54	33.96
12	29.36	30.80
15	26.30	27.82
20	22.50	25.12

$200 loan on which capital is repaid in equal annual instalments is shown. On a $200 loan over 10 years an easing of the interest rate by 1% from 11% to 10% reduces the maximum annual repayment by only $1.42. But by extending the 11% loan from 10 to 15 years, the maximum annual repayment is reduced by $6.14.

Lengthening the period of the loan reduces the annual repayment required, but is helpful only up to a certain point. As the length of the loan increases, the reduction in the annual repayments becomes less and less because the interest component of total repayments increases. So while lengthening the period of a loan initially reduces the annual repayment required, it is only helpful up to a certain length of loan.

Practice

If a farmer could increase operating profit by adopting technology and/or by changing enterprise mix, the following points will have a large bearing on the outcome of any programme undertaken.

Normally, no substantial increase in operating profit will occur unless extra capital is injected into the farm through bush clearing, better equipment and buildings, fertiliser, crops, water supplies. This extra capital can sometimes be financed from current operating profit, but usually has to come from outside sources.

The return on extra capital invested for intensification can be as high as 25% after tax; 20% is quite a common figure. Because of risk, it is not usually desirable to invest in development unless the after tax return exceeds 10–15%. Lenders are generally not prepared to advance funds to projects which cannot be paid off in 10–15 years. Eight years is a common requirement, but it is usually too short a time for most agricultural investment projects to repay the capital borrowed.

Tax and interest have to be deducted from any increased operating profit before the extra funds earned can be directed towards reducing the principal of the loan. The level of expected increase in the operating profit, less tax and extra interest, rather than the equity of the proprietor, will determine the project's ability to service the loan. Financiers do not want to see equity drop too low, as this means a rise in the interest payments and makes repayment difficult in the event of a drought, an outbreak of disease or a marked decline in prices.

Repayment plans need to take account of the possibility of unfavourable seasons and prices, and of negative cash flows in the early years. It is not easy to prepare a good budget embracing a long-term (5–8 year) development plan, especially where sound decisions on complex technical matters involving crops, stock, plant and buildings, are essential to the success of the plan.

When the broad outline of the plan and the tentative development budget has been prepared, it is wise to have a 3 or 4 person conference of those involved (e.g. banker, adviser, farmer, merchant) to discuss it, with a view to refining it. It is virtually impossible for a single 'expert' alone to draw up a plan which is likely to be successful, feasible and, at the same time, acceptable to both farmer and financiers.

Such a conference will allow the interested parties to recognise the framework in which the farmer is working, and to set yardsticks by which progress can be measured. Inevitably, the programme will have to be modified as time passes, but it is more likely to be successful if the financiers know, at the annual review of progress which has been made, that their client is working to a systematic plan.

A farmer's ability to earn growth profits often depends on obtaining credit to purchase new inputs, to replace items of equipment or to even out uneven cash flows. The uncertainty of returns in farming activities means there is no certainty of credit adding to profits or being repaid. Farmers need and use credit in situations characterised by high risk, great variability of returns and often precarious equity/debt positions. The critical question of 'to borrow or not to borrow, to lend or not to lend' has to be answered with a good deal of thought and care.

The answer to the question 'how much to borrow' depends upon borrowing capacity, repaying capacity and risk-bearing capacity. Borrowing capacity depends considerably on equity position; repaying capacity depends on annual cash returns; and risk-bearing capacity depends upon equity ratio, the principle of increasing risk and individual attitude to risk.

The most important requirement when applying to

Table 15.2. *A checklist of items for an annual loan appraisal*[a]

	1st season	2nd season
Cash receipts		
Crops		
Animals		
Animal products (milk, hair)		
Off-farm work and trading		
Money from family members		
Sale of assets		
Other cash income		
A. Total cash receipts		
Cash payments		
Living costs		
Schooling		
Farm inputs (seed, fertiliser)		
Taxes		
Gifts and tributes		
Repairs to buildings, machines, plant		
Travel		
Other cash payments		
B. Total cash payments		
C. Cash deficit or surplus (A − B)		
D. Loan needed		
E. Interest and loan repayment		

[a] For an *annual* loan, data on the following items are needed by both borrower and would-be lender.

borrow farm finance is do the homework. That is, to have the adviser present a sound, well-documented case to the prospective lender. When financing farm development or expansion, the ability to service debt and equity are the major considerations. Ability to service debt depends on the level of the annual cash surplus before deducting interest and principal repayments. Equity is the net amount which would be realised if the farm was sold and all debts paid. The equity to borrowings ratio is a useful guide to the safety of a borrowing programme.

A set of questions is given in Tables 15.2 and 15.3, for two lending situations, viz: annual loans, and loans for development (more than 3–4 years). The questions relate mostly to the lender, to the case where a bank, or government body, or cooperative is lending to a farmer. The checklist is primarily designed for use by a loans officer working for such organisations. The list of questions also indicates aspects of the proposed loan which the potential borrower needs to consider.

Questions

1 Of the two broad groups of farmers requiring credit cited on p. 146, into which category do the farmers you know fit?

2 Where do these farmers get credit? What does it cost them? What levels of equity do they have in their businesses? What is the lowest level of equity which you consider to be safe?

3 How is debt-servicing ability calculated?

4 Explain the link between leverage and growth of equity.

5 What does the principle of increasing risk say?

6 What determines the growth of a business?

7 What is the difference between a market rate of interest and a real rate of interest?

8 $m = r + f + rf$. What does this mean?

9 What is the difference between flat and simple interest?

10 What are the major determinants of borrowing capacity, repaying capacity and risk-bearing capacity?

11 What are the main pieces of information which you will need in order to prepare a case for (i) an annual loan and (ii) a development loan?

Table 15.3. *A checklist of items for a development loan appraisal*

When considering a loan for development, the lender needs to be able to put reasonably accurate estimates into the following tables:

1. Value of land owned
    Value of plant and equipment
    Value of land improvement
    Value of stock owned
    Money owned
    Money owed by others
    Total (*A*)
    Existing loans owing
    Other debts
    Total (*B*)
    Equity (*A* − *B*)

2. For farm plan in steady state
    Gross cash income
    less Cash operating costs
        Living expenses
        Plant replacement costs
        Capital investment costs from cash,
            not loans
        Taxes
    Annual cash surplus
    Will this amount cover interest and
    principal charges?

3. Cash flows   *Year 1 Year 2 Year 3 Year 4 Steady state*
    (*A*) Cash in (including loans)
    (*B*) Cash out (not including interest and
        principal)
    (*C*) Annual deficit/surplus before
        principal and interest (*A* − *B*)
    (*D*) Annual debt servicing costs
        (principal, interest, HP repayments)
    (*E*) Annual deficit/surplus after debt
        servicing (*C* − *D*)
    (*F*) Cumulative deficit or surplus

4. Overview
    Total borrowings (new and old loans)
    Present level of equity
    New equity if given loan
    Annual debt servicing costs
    Expected steady state annual surplus
    Timing of peak debt
    Time until cumulative cash flow is in
        surplus
    Time needed for loan repayment

16

Beyond the farm

Farm management is primarily about things which happen on the farm. The things which happen on the farm also depend strongly on what happens beyond the boundaries of the farm. Decisions taken by governments about pricing, marketing, taxation, health, welfare services, credit, land tenure, education etc. all profoundly and directly affect what can be, and is, done on the farm. Extension officers, agricultural students, farm managers and rural financiers have to take positions on, and respond in the field to, the rules and to the changes in the rules made by governments. Awareness of the economic forces at work beyond the farm is an integral part of good farm management.

We noted in the introduction to this volume that many tropical countries lack capital, have low levels of savings and capital formation, and low incomes per head. Most of their populations work in agriculture, which contributes a large proportion of the national income. Most countries can earn export income only by exporting rural products in a relatively 'raw' form. As well, much agriculture is either fully or semi-subsistence. Population pressure often causes massive degradation of the natural resource base. Protein intake per head, health, living and educational standards are generally low.

There is nothing so far in this volume about what the government should do. We intend this volume for local and regional extension officers advising small farmers, the managers/operators of medium-sized commercial farms, people in an agricultural teaching role, students, lending officers, cooperative managers, and local or regional planners. The volume deals with how to apply principles of farm management to get 'a bit more' from the limited physical and financial resources available. Its prime targets are those working in the field and the classroom, not in air-conditioned offices. We have said nothing about the way each region and country tackles its unique set of problems of how to get 'a bit more' for its people. There are two good reasons for this:

(i) understanding the causes of economic development is a specialised field of study, which is not our forte;

(ii) there is no one path to 'pastures of plenty', and there are always several equally valid ways of attempting to solve economic problems.

Even for specialists who study the economics of development, and also for those who may agree on certain approaches 'in principle', each region or country is unique. Each situation can only sensibly be discussed after detailed case studies for each 'area' have been made. Our aim here is to record some general observations of some things that do not always seem to work very well. The following is a selection of our thoughts about actions/priorities/methods of getting things done.* Each needs very careful investigation before it is transformed into a slogan which becomes the 'truth' or the 'way'. Let us stress that we do not believe that there is any one best solution to any particular problem, and that we regard the different ways of organising societies as essentially different ways of tackling quite similar basic problems.

Causes and effects: a note

With economic processes, the immediately obvious causes and effects of an action are not necessarily the

*'Our' thoughts include many ideas distilled from many and varied perspicacious, enlightened, and enlightening, sources – including Dr Alistair Watson, School of Agriculture, University of Melbourne, Australia.

whole, or necessarily the main, part of the story. The eventual outcomes of initial actions might be the complete opposite to the effects which were intended or expected.

Economists worthy of the title look through the first round of causes and effects, and then consider the second-, third- and even later-stage effects of a change. For example, introducing new plant varieties, new or more fertilisers, different cropping rotations, all may have the objective of improving the lot of a poor rural population. They can, however, mean that most of the benefits go to only a minority who have sufficiently large farms and access to the credit, relevant technology and information necessary to produce a surplus. The worst off might be made even worse off.

Alternatively, the introduction of machinery may save labour, increase the productivity of land and labour, increase food output while causing a growing number of people to be unemployed and landless. These 'landless poor' may now be even less able to buy the food which is being produced than they could previously, when less food was being produced. The adoption of new techniques may improve the economic position, borrowing power and the command over resources of some farmers but have the opposite effect on many others. Particular causes are sometimes incorrectly blamed for subsequent effects. For example, a country might embark upon a programme of agricultural development involving a range of strategies. New techniques, new varieties and new varieties adopted, and more purchased inputs used. As well, government might set up a series of parastatals and marketing boards to produce some products and to market outputs and inputs. If the expected agricultural leap forward doesn't materialise, this may be because (i) the new market-based technologies were inappropriate and/or (ii) the government bodies set the price signals wrongly, thereby precipitating shortages of agricultural outputs. The latter cause is less obvious than the former, hence blame may be unfairly attributed.

Analysis of a sequence of economic causes and effects may reveal that a programme, albeit apparently beneficial, may in practice make it more likely that the situation will become worse than if no action had been taken at all. So, the more that possible second-, third- (and later-) stage effects are considered in policy decisions, the better. Conclusions about possible outcomes of an action should be arrived at by understanding the key linkages between causes and effects. If these are both done, then seemingly good ideas may actually benefit those people the ideas are intended to help.

The market and government

There are many means by which competing interests within an economy are reconciled. There are elements of the operation of markets and of government regulations in all economies – it is the mixture of market and government, the natural endowment and ownership of resources, and the objectives, which vary. Decisions have to be made about the extent to which the government is involved in providing and overseeing the incentives for (and the pattern of) resource use in agriculture, and also on the degree to which the incentives to allocate and reward resources are determined by markets and prices.

To what extent, or in which areas, do markets act to allocate resources in a socially inexpensive, beneficial and effective way; to what extent, and where, do markets fail? Identifying where markets work and where they fail, and judgments about how well they work or how badly they fail, are at the heart of decisions about running an economy.

Whatever is the preferred system of economic organisation, it is crucial that the incentives and rewards for human endeavour be compatible with goals of efficiency and equity; facilities for converting production surplus into capital formation, whether for public or private investment, should be well established. This requires that economic information is documented and analysed. The other components of society (such as laws, customs, systems of land tenure, traditional attitudes to property, religion etc.) have to be considered. For instance, land reform is occasionally considered a cure-all solution to problems of poverty and low productivity. The form of land tenure can reduce or increase the incentives to produce surpluses and the potential to improve production.

The issues involved in the day-to-day running of an economy are often far removed from another area of debate which is equally important. This area includes the fundamental philosophical debate about the nature of how an economy works and how it should work, about the meaning and sources of 'value', and about the distribution of the rewards from production.

Everyone knows that whenever two or more people gather, there is potential for initiating agreements of exchange, for establishing prices for goods and services. Most people know that there is also potential for conspiracy to extort higher than competitive market prices from buyers, or force lower than competitive prices upon sellers. Where there is a market with lots of producers competing to sell their products, and many buyers competing to buy the products, the consumers' demand for the goods will tend to be met by producers making their goods as cheaply as they can so that they can sell their output.* It needs a lot of information for this to occur, and the price-establishing processes in a busy market itself generates an enormous amount of information about what consumers want, how much they want, how much they are prepared to pay for it. As well, it tells the producer how well he is using the resources available to him in comparison with his fellow, competing producers.

This simple story, like all simple stories, cannot be extended across a whole economy, but there are some truths for particular cases and particular products. A competitive market, where it exists, can be a very powerful generator of information and provide valuable guides and incentives to using resources efficiently. Even when administrative devices are used for making production and consumption decisions, the market eventually has its say. If a marketing board sets its price too high, stocks will accumulate and waste, or the price will have to be lowered to clear them. Similarly, if the price is too low, producers will make something else and stocks will have to be rationed in some way (e.g. by people queuing or by allowances and quotas per person), or the price will have to rise to ration the limited stocks. Again the market is having its 'say', albeit indirectly.

Marketing boards and markets

Government involvement in production and marketing decisions is often manifest in the establishment of marketing boards. Marketing boards can have a

*The ability of consumers to demand goods and services depends partly on the distribution of income which exists; that is another, very important, issue.

constructive role to play in correcting market failures, for instance, by balancing the market power of a large buyer or seller in the market, or by ensuring that the necessary quality and quantity of a product is supplied consistently to consumers. In fulfilling these valuable roles, however, boards are sometimes open to criticisms. They fail to adjust to changed circumstances and the seeming irreversibility of their role can lead to excessive costs and inefficiencies.

This need not be so. A board competing actively and efficiently with other marketing intermediaries can place both producer and consumer in a position where they cannot lose. If a public marketing board can perform better than its private competitors, say because of its scale of operations, then producers and consumers will benefit. If it cannot carry out the marketing functions as well as properly taxed private competitors, then the private sector will be providing the service instead of the board. Again producers and consumers stand to be better off than if they were to be provided with these services by the less efficient marketing board.

Marketing boards may be established to control supply of a staple or export foodstuff. The success of this venture depends on how well the price-setting procedure balances the competing interests of producers (who want a high price), and consumers (who want a low price). It is difficult to get it right, and getting it wrong can be costly.

Marketing boards sometimes act as devices to channel consumer funds to producers. Even if this is not the case and their objective is to set prices of products so that (i) both producers and consumers are satisfied, and (ii) all are being told the right 'messages' about what to produce, how much and how, the task of generating by administrative means the same amount of information as the market system does, is extraordinarily difficult.

A board may also exist to provide cheap staples to many poor and hungry consumers, and so the price is kept low. Many consumers may be made better off by this but there are other possible effects – somehow producers have to be encouraged to produce the food, and if they are paid only a small price they will either not produce it, or not supply it to the board, and instead sell it on the then (usually) burgeoning black market. The net result can quite easily be that those whom it was intended to benefit end up being worse off, or some other group ends up worse off, because somehow the government has to find the funds to subsidise the producers, to ensure that they supply the needed goods. The intended result not only does not happen, but can even set in train a series of events which guarantees that the opposite outcome will occur.

Other problems peculiar to the nature of boards (apart from difficulties related to the inherent problems of planning) are that, unless a board can control all the supply, there are producers who do not contribute to finance the board's operations but who also stand to gain from it (called 'free riders'). Also, there can be strong incentives at times for producers to dispose of their product outside the board. In fact, the more success a board has in restricting supply and raising prices, the greater the incentive to sell outside the board. This creates problems of enforcement, and difficulties in maintaining 'artificial' prices.

There are better ways of tackling fairness, or equity, objectives than by interfering with the mechanisms of food market prices. Pursuing 'equity' objectives by interfering with what is essentially an efficiency mechanism can cause inequities and inefficiencies. If resources are used well, then the cake which we want to divide evenly and fairly may well be bigger.

The nature of a product, and the conditions under which it is supplied and demanded, determines how suitable it is to be marketed through a board. Boards are not the 'solution' to problems of production and marketing in developing countries, any more than they have been in the more developed countries. Do not think that we have ignored considerations of equity. It is just that there are often more effective and efficient means of redistributing wealth than through the operation of marketing boards.

Government in farming

Often there are government and semi-government attempts to provide services which could probably be more effectively supplied by an appropriately regulated and taxed private sector. There are numerous examples of economic disasters and near-disasters which have occurred in such activities as federal, state, local government and parastatal farming ventures, tractor-hiring services, marketing boards, distribution of inputs, the import and supply of foodstuffs to consumers and sometimes government-backed cooperatives.

A build-up of bureaucracies with exceptionally low levels of productivity, and resistance to measures based on such criteria as 'profitability' and sound resource use is all too common. Equally important, excess government participation in practical agriculture can create the situation whereby responsibility and accountability are not fostered. It is called the 'it's the government's money, so it doesn't matter' syndrome. It is often as though economic analysis on the use of government funds is something which cannot be done; and to be accountable to the providers of those funds would be too burdensome an imposition upon the spenders. This certainly is not confined to tropical countries – it occurs in many places throughout the world. The following letter to the editor of the *New Nigerian*, published in 1981, exemplifies the simplistic thinking of which we are critical.

Kindly permit me a space in your newspaper to submit the following suggestions to the Federal Government on how best to make the 'Green Revolution' programme a success. Below are the suggestions:

Sufficient funds should be ploughed into the programme to make it a success.

River basin authorities should be set up in all the states of the federation.

The Federal Government should acquire sufficient area of land in all the states for the programme.

The Federal Government should employ the services of foreign experts from both the Western and Eastern blocs where mechanised farming has proved successful. These experts from the West and East should not be mixed. They should be deployed to different tasks and assessed at the expiration of the deadline stipulation for them.

Credit

Where the financial reserves for credit are going to come from in the first place is a very real problem at the national level; it depends on the nation's production, consumption and savings, export sales and import levels, and overseas borrowings. Though this is true, it is not our concern here.

In some situations there is a role for cooperatives as credit institutions, supplying small amounts of closely supervised credit as well as, at the same time, educating farmers to make best use of such credit.

Providing credit at cheaper than market rates is not usually the good thing it may seem. Cheap credit policies for agriculture can ration credit away from agriculture and into more profitable (for the lenders) use. As well, cheap credit can discourage savers and financial institutions from involving themselves with agricultural credit and can lead to less efficient use of resources (not just the credit), than would otherwise occur. It can actually worsen already inequitable distribution of income. Also, it can foster the use of more capital inputs (e.g. tractors) relative to labour inputs. Even for those who get cheap loans, they may find they cannot get as much as is needed because it is rationed, and may be forced to borrow the rest from an exorbitantly priced source of credit. Again, it is a case of seemingly obvious solutions having the wrong effect.

Greater flexibility in loan repayment terms, plus supervision and education in planning the use of credit, seem a fundamentally better approach to any problems of credit shortage which small farmers may face. Credit can, if wisely used, help small farmers to increase productivity and output. Because of risk, management and debt repayment problems, greater flexibility in the timing of loan repayment is much more likely to be of real help to borrowers than are concessional interest rates. It seems obvious that people making decisions about lending money to farmers need to have a good grasp of the technical as well as the economic aspects of the farming activities into which the funds are to be put. Often, in practice, this is not the case.

Mechanisation

At the most basic level, the debates about mechanisation (particularly the use of tractors) in small farming, focus on the following two opposing lines of thought:

(i) increased mechanisation involves using a scarce factor of production (capital) to replace a plentiful factor of production (labour), thereby causing further unemployment and underemployment, with undesirable social consequences;

(ii) it is the total productivity of the farming sector which determines how well off rural populations

can be, and so the use of new technology, such as machinery to increase production, is essential (continuing labour-intensive, non-mechanised agriculture will tend to maintain rural poverty and stagnation, at least once the effects of other, non-mechanical, new inputs reach a certain level).

In our opinion, the answer depends on case studies of particular circumstances.

Hiring the services of a small tractor for the major tillage operations might not displace much labour, and may be essential to attaining a more productive, more intensified, yet still timely, plan of cropping operations. On the other hand, introducing a machine for a labour-intensive, but not labour-bottlenecked, farming operation may have serious economic and social consequences for a rural work force.

Productivity is usually low, holdings are often both small and in scattered plots. This results in a poor standard of living for most of the rural population. Most crops are cultivated by non-mechanical means. Without going too far, low productivity can sometimes be raised through new technology and new methods – part of which involves mechanisation of some tasks. A problem might be that while some mechanisation is needed to increase crop production, the appropriate machinery might not be available. Also, the system of land tenure leading to scattered plots may make machine use impractical.

As new seeds, fertilisers, plant varieties, crop combinations and crop rotations increase yields, small farmers need more frequent, and more timely cultivations. Thus some mechanisation of cultivation (which does not necessarily mean individual ownership of machinery) can have a role to play if there is no serious problem of land fragmentation. Mechanised land preparation can improve soil preparation, timeliness of operations and thus yields, allow more intensive cropping, or increased crop area where labour was the limiting factor; it also can reduce drudgery.

Intensified cropping increases the number of tillage operations and the importance of timeliness. A real need is for appropriate-sized power plants and implements – which are of simple design and robust. Simple changes and improvements to existing machines, or simple pieces of 'minor' equipment other than for primary tillage, can be very useful, e.g. rubber tyres, traditional wooden tools made of light strong materials, cheap, simple water pumps in a range of sizes. Mechanisation creates a vital need for back-up replacement parts, for repair and maintenance skills and local service skills. Machinery purchase, repair and operating costs are often high, and can include a burden on balance of payments.

In some areas hand cultivation will continue to be the best, or only, method possible. The use of mechanised, powered equipment can mainly be justified if it overcomes labour shortage at critical times or improves the quality of operations. It may be that the elimination of a labour bottleneck might allow more complete use of labour over the entire season rather than causing any reduction in the total amount of labour used. Seldom will individual farms be large enough to make profitable use of a tractor and implements. Mostly, mechanisation is a realistic proposition only where contract services are provided for hire to farmers who themselves cannot afford (or economically justify) owning their own equipment, where land in scattered parcels can be pooled for mechanised cultivation, or where large state or collective farms exist.

The problems of general rural poverty are not reduced if machinery is used to achieve the same output using less labour – what is needed is to use machinery and other new technology to increase total production. Planners have to know the particular capital and labour needs of particular groups of farmers in order to make informed, case-by-case judgments on the likely consequences of different degrees of mechanisation. We also think that it is very important that planners in developing countries recognise, and learn from, the many harmful economic and environmental effects caused by policies which encourage the use of private cars at the expense of public transport in developed countries.

Education

What role has formal education in the growth of the farming sector in developing countries? Some say the more the better – education is a good thing, it is also a basic right. This is an ideal, rather than a prescription for action. Others say that, with low levels of education for most of the population and with limited resources, the greatest benefit to society comes from exposing a lot (of the population) to a little (education). For example, 4–5 years schooling, some rudimentary reading, writing and numeracy skills. Still others advocate policies best described as a lot (of education) for a little (number of people).

Desire for greater knowledge for knowledge's sake, for material betterment, for prestige, to satisfy curiosity, to do something better, or to help others are all apparent motives in the pursuit of education. To know, to understand and to apply some of these ideas ranks in the list of human wants (p. 00) which we presented in the introduction to this volume. Information and knowledge can come from the 'words of the wise' such as elders and local leaders, from traditional stories, ballads and verse and via the radio and television. The written word is also a valuable source, and this means that it is 'good' that people should be able to read to explore ideas further.

Education raises people's expectations. Raising expectations can create problems when expectations are then not fulfilled. Whilst this cannot be put forward as an argument against education, it is valid to note that the products of the secondary school system frequently do not want to return to the village, as their expectations have been raised above the occupational opportunities and lifestyle which the village offers. Unfortunately there are often not enough jobs for them, and they can be neither good farmers nor sufficiently qualified to find good jobs in the urban sector. Many university graduates in tropical countries believe that possessing a degree entitles them to wear a suit and tie for the rest of their lives, never to have to get their shoes muddy, and to be paid well, regardless of their performance. It is common to see universities producing large numbers of people trained in subjects which are of less immediate value in a developing country than more applied training; for example, in 1980 one university in Africa produced 200 political 'scientists' and three agricultural engineers!

We showed with our weeding example in Chapter 4 (p. 00) that economic techniques can sometimes be useful in helping people to decide how best to allocate scarce resources between (competing) alternatives. The central concept we stressed was the need to assess how much extra return came from adding one more unit of scarce input. This concept can be applied to decisions on how best to allocate the scarce 'education dollar'. Certainly it is more complicated, but it is still desirable to try to assess what educational activities give the best return to the economy for the scarce dollars used.

Training for farm management

The small farmer who operates mostly in a subsistence, non-money economy, and only partly in a cash economy, may not respond to market 'signals' to the same extent as the farmer who is more 'commercial'. Making decisions, even in a semi-subsistence economy, is still something which the small farmer has to do. Questions about what to produce, and how, have to be answered. Selected principles of farm management apply, even here.

Most semi-subsistence farmers eagerly adopt relevant new technology. They are often hampered in their efforts to improve their performance by poor input-distribution systems. Frequently, lack of advice from sources which they deem to be credible also stops them from getting the best from their resources. Too often there is not enough effort made by governments (at all levels) to remedy these problems. Also, a disproportionate amount of funds in tropical countries is often invested in elaborate, ineffective physical structures, relative to that spent on manpower development to service the agricultural sector.

One issue at debate in agricultural development regards the form of farm structure and organisation which should be encouraged. There are fundamentalists who abhor any intervention with the idealised traditional semi-subsistence farm structure and land use. Opposing them are advocates of big machine, large-scale, high technology, high chemical-input farming. Neither party holds a monopoly of the truth.

First, let us recognise that most farmers are clever people. It is essential to understand the farmer's situation and thinking, and not to give simplistic prescriptions. At best they are useless, at worst damaging. Second, it is apparent that most traditional farmers accept and adopt new appropriate agricultural technology if they can, in whatever form it might take, simply because they make more money by doing so. Third, there is scope for some farmers to use a small amount of mechanisation (usually hired, rather than

owned) to provide the timeliness of operations which is so critical in obtaining good yields. Fourth, there are diseconomies of scale in large-area farming, so these big ventures rarely work, relative to the results which medium-sized and small, efficient, well-managed farms can achieve. Fifth, there is a biological limit to exploitive land use. Excess reliance on chemicals can also be self-defeating after a certain point. We hope that the accumulated experience of traditional farmers, combined with the technical knowledge of the extension worker, can prevent such situations from becoming too damaging.

We consider it best to encourage both efficient, small, semi-subsistence farmers and medium-sized, but not large-sized commercial operations (be they semi-government or private). This will mean that many more people trained in farm management are needed, both to advise small farmers and to help manage medium-sized farms. In most tropical countries there is too little investment in manpower training at all levels in applied farm management economics. Few institutions offer training in applied farm management at the operational, as distinct from the academic, level. There is a severe gap between demand for, and supply of, competent people trained in farm management. How can you have an effective 'Green Revolution' (or any other sort of revolution for that matter) when you do not have good people in the field to manage it? We agree with Dr Onazi of the Division of Agricultural Colleges, Samaru, Nigeria, when he says:

I am convinced that a cardinal principle in planning is: 'start with the known'. This implies, for farm management training, that we improve the existing institutions who have any connection with agricultural training at the applied, not theoretical, level. The way to improve them is to introduce a solid farm management component into the training.

It is vital to get the most effective use of the money spent on development and also to achieve the objectives of programmes. So it is necessary to have a proper balance between the money spent on physical structures and that spent on training and properly rewarding the people needed to carry out the programmes. There is a common fault made by planners and governments in tropical countries where rapid development is being pursued. They spend too little on skilled manpower, especially on practical farm management, trades and skilled field operators. A result is that much of the investment in physical resources is not put to best use; at worst, such investment hardly achieves the objectives of the programme at all.

It is people, not structures, which need to be 'built'. A shortage of people trained at the field operational level – builders, mechanics, plant operators, crop and animal husbandry practitioners, technicians – severely hinders nearly all development programmes. Such vocations have relatively low status compared with those followed by university graduates, even though their immediate contribution to the economy of the country may be much greater.

N.R. Carpenter, in reviewing education for agriculture in developing countries,* said that for every university graduate there should be at least 30–40 people trained at the level of the agricultural college field technician. Our experience would confirm what he said; this volume is written partly for them.

We stressed in earlier chapters that money is a scarce resource and has many alternative, competing uses. Investment in education is directly expensive for the state, especially at the secondary and tertiary level, and is ultimately even more costly if it is not very effective.

As an example of how *existing* structures and institutions (rather than new ones) could be used, we list some of the key proposals which J. P. Makeham made for the involvement of Nigeria's Admadu Bello University (College of Agriculture). He suggested using the facilities during college vacations to expand farm management training to many more people. We stress that this is just an example. Many other institutions in the tropics already have the necessary infrastructure to follow a similar approach. The proposals are:

short (8–10 weeks) courses are to be the basis of the programme;

the method of teaching to be based on actual farm case studies, applying the general principles of farm management economics and agricultural science;

use the institutions when the full time students are on vacation.

*N. R. Carpenter *Priorities for Education in Developing Countries.* Paper presented to FAO conference on rural education, Rome, 1980.

The following types of people would be trained at these short courses:

(i) extension workers who will be used to give farm management advice to operators of small- and medium-sized farms;

(ii) practising managers and sub-managers of government and private farms;

(iii) farm management advisory officers of existing large projects;

(iv) teachers of agriculture who have had little farm management training and who will be teaching introductory farm management courses in colleges and related institutions;

(v) loan-appraisal officers of private and semi-government banks and cooperatives who are involved in lending to farmers;

(vi) other persons from either the private or the government sector who are dealing with farmers at the applied farm management level.

Knowledge and skills to be taught during the intensive 8–10 week course

Note: teaching would be based on case studies.

Knowledge
The following list covers the sort of information about farm management which needs to be known and understood by people who are closely involved in agricultural development, and/or are dealing with and advising farmers.

(i) how human, physical, financial and off-farm factors interact to produce output and income from any farm or farming system;

(ii) what factors can be controlled and/or directed by people, and what factors cannot be;

(iii) the extent to which it is possible, in practice, for people to control and/or direct each of the main factors;

(iv) the principles which determine whether a farmer will adopt or reject an innovation (a new practice);

(v) economic principles and techniques which show how to choose the most profitable enterprises and/or mixture of farm enterprises;

(vi) techniques for making farm management decisions when the amount of data (information) is incomplete, and when the outcome of a plan is quite uncertain.

Skills

The following are the skills, related to the points listed in the previous section, which the student should possess at the end of an introductory farm management training course of this type.

(i) (a) Define the farmer's strengths, weaknesses, skills, interests and objectives for himself and his family. Determine the skills, amount and type of labour available (own, family, hired, shared) to carry out farm operations correctly, on time, at critical times of the farming year;

(b) analyse the physical resources (soil types, water, buildings, animals, machinery range, land) available;

(c) define the financial structure of the farm:

form of land tenure;

value of farm land, buildings and other farm assets;

amount and type of debt (short, medium and long term), plus annual loan repayments and interest;

(d) identify other factors which influence profitability of the farm. These include marketing, transport, storage, subsidies, existence of a cooperative.

(ii) List the main factors which can be controlled and/or directed by the farm operator/manager on the particular farm being studied; these will include: choice and mix of enterprises, type and level of input use, control of some disease, weeds and insects, labour, type of cultivation, quality of seed-bed, type of power used, drainage.

(iii) Determine the *extent* to which control and/or direction can be exercised. For example, the soil type may restrict the farmer to a choice of 2–3 crops, the weather may affect yield, even though the correct inputs (seeds, fertiliser and spray) have been used, and so on.

(iv) Determine what factors a particular farmer considers before he will adopt a new practice. For example, what local evidence is there that the new practice will work? What are the risks and the expected benefits? Is there a lot of extra work or new skills needed for it to be a success? Are the costs too high? Is it a simple change from existing practice, or is it complex? How can he do a small trial on his own farm? What will his neighbours in the village think about him doing it?

(v) Carry out the following main forms of economic analyses and budget:

enterprise gross margins;

whole farm overhead costs (including cash living expenses and taxes);

annual debt repayments;

whole farm economic analysis;

partial budgets to assess the likely effect of introducing a new enterprise or mix of enterprises;

capital budget;

household food demand and farm food supply analysis and budget;

cash flow budget;

finance budget;

whole farm budget for the next year.

(vi) In order to help cope with the problems of uncertainty and risk when he is advising on planning for improving the profitability and/or the standard of living of a farm and the farmer, the student must be able to:

obtain the farmer's attitude to risk, and the probability which he (the farmer) assigns to the possible outcomes (good, average, poor);

attach money values to each of these outcomes;

calculate the effect on the total farm profit and standard of living if the outcome is worse or better than expected;

help the farmer to draw up contingency plans in the event of both the (i) worse or (ii) better than expected, outcome;

help the farmer to decide on a plan aimed at increasing profits but keeping the harmful effects of risk within limits acceptable to the farmer.

The telecommunications revolution

Developments in satellite technology and electronics over the next decade may make it possible for people in rural villages to be provided with much more relevant information, at low cost, than ever before. Also such developments may reduce the need for travel, which is presently needed for 'in person' communication.

The era of being able to readily phone someone and also see him/her on a video screen whilst you are talking, is coming. The potential savings in time and money are huge. An article (December, 1983) in the *Economist*, titled 'The capitalist luxury that is a Third World essential' makes the following points on these developments:

The developing world has been demanding the 'right to communicate' but without the means to buy even the most basic technology. About 15 per cent of the world's countries have 90 per cent of its telephones and many of the rest have little hope of joining the international network for another 50 years.

This month the United Nations' World Communications Year draws to a close with the setting up of a commission and a fund to try to spread telephones faster. The commission seems set to recommend high technology as a cure for the developing world's telecommunications ills. . . .

A general axiom behind the commission's thinking is that for countries who are late-comers, *the appropriate technology is often the most sophisticated technology*. Communications satellites, especially, provide the chance to overcome the lack of an existing telephone system as their signals vault over rough, wet or empty terrain. But optical-fibre wires, digital telephone exchanges and mobile radio also make it possible to have a working up-to-the-minute telephone and data service where there was none before. . . .

The World Bank has long put telecommunications at the bottom of its priorities for aid. But so too have many politicians within the countries in need, because of three widely held beliefs:

– that telephones are a luxury which the poor do not want;
– that telecommunications are so profitable in urban areas that even poor countries can finance expansion from their own resources, and
– that investment in telecommunications does little to stimulate economic growth or to help the poor. The World Bank has tried to channel its limited aid money into projects which will improve the lives of the very poorest.

These beliefs may be misplaced. The potential contribution of improved telecommunications to making the lives of the rural poor better should not be underestimated. For instance, the *Economist* article continues:

In the rural regions of Andhra Pradesh in India, an OECD study by Mr S. N. Kaul showed that when a community telephone was available in a farming village, even the poorest people were willing to pay a hefty share of their month's income to use it. Had there been no telephone for urgent calls, they would have carried the message in person, by bus, even at the loss of several days' wages. The villagers' biggest complaint about their telephones was the *unreliability* of the service – not the expense.

Telecommunications, along with power and efficient transport, is one of the pillars of the infrastructure on which improvements in social and economic life can be built.

A major impediment to getting the potential gains from improved telecommunications, (once the more basic infrastructures such as reasonable transport and reliable power are provided), is how to maintain the system. If there are not enough *skilled, field level technicians* to keep communications systems running reliably, and no efficient organisation to support their work, then the hoped-for telecommunications 'revolution' will more likely be a telecommunications 'hallucination'.

To conclude

Perhaps the cumulative effects of the inventiveness and the perseverance of people worrying away at the big problems facing tropical farmers will lead to gradual improvements – in yields, quality of agricultural products and productivity, savings and investment, as well as decreases in rate of population growth, and improvements in the quality of life of the rural poor. The best that can be hoped for is 'a bit more', probably not a lot more.

We hope we have done our 'bit'.

Appendix 1: Interest rate tables

The following tables come from the *Professional Farm Management Guidebook No. 2, Discounting and Other Interest Rate Procedures in Farm Management* by A. H. Chisholm and J. L. Dillon, published in 1971 by the Department of Agricultural Economics and Business Management of the University of New England, Armidale. We are greatly indebted to these authors for their kind permission to use them. The derivation and use of these tables is fully explained in Guidebook No. 2, which is one of an excellent series of farm management booklets from the Department of Agricultural Economics and the Agricultural Business Research Institute, University of New England, Armidale, NSW, Australia. To help you understand what may appear to be a bewildering array of numbers, the following explanations are given here.

In all of the tables, the number of years is shown in the first column at the left hand side of the page, headed 'Years', and 'n', which means '$number$ of years'. The interest rate being considered is shown in the row at the top, e.g. 0.01 (1%), 0.10 (10%). In the body of each table is a factor appropriate to the number of years and the chosen interest rate. For example, in Table A, the present value of one dollar in 20 years' time, at an interest rate of 5%, is $0.3769 (roughly 38 cents).

All of the tables are based on the assumption that $1 (or 1 unit of any currency) is the item being considered, e.g. in Table A, the future lump sum is $1. In Table B, the factors in the body of the table tell you how much $1 will grow to if invested at compound interest for a given number of years at a particular interest rate. Thus, with an interest rate of 5%, $1 invested now, with the interest it earns re-invested each year, will grow to $2.65 in 20 years time (see Column .05 (5%), *row* 20 years). Where the amounts considered are more than $1 it is a simple case of multiplying the factors in the table by the sum involved. For example, $10 000 will grow to $26 500 (10 000 × 2.65) if invested for 20 years at 5%.

We will now examine each table in detail. In Table A, the current value is shown of $1 received at some time in the future, at specified interest rates. Table B presents the factors by which $1 will grow if invested at compound interest.

An annuity is an equal sum of money received or spent over a period. Thus, in Table C, the factors in the body are today's value of $1 received or spent each year for the stated number of years and interest rate. If you were to get $1 each year for 20 years, and the relevant interest rate was 5%, then today's value of this flow of $1 is $12.46. In Table D, the value to which $1 received each year will grow, if the interest from it is also reinvested, is shown. For example, $1 per year at 5% interest for 20 years becomes $33.06.

In Table E, we show the *parts* of a dollar which need to be received each year in order to grow to be $1 at the end of the period, e.g. $0.0303 (or 3 cents) received each year for 20 years and invested at 5% interest will grow to $1 after 20 years.

Table A. *Present value of a future lump sum*

Years Interest rate (i)

(n)	0.01(1%)	0.02(2%)	0.03(3%)	0.05(5%)	0.08(8%)	0.10(10%)	0.12(12%)	0.15(15%)	0.18(18%)	0.20(20%)	0.25(25%)	0.30(30%)	0.35(35%)	0.40(40%)	0.50(50%)
1	0.9901	0.9804	0.9709	0.9524	0.9259	0.9091	0.8929	0.8696	0.8475	0.8333	0.8000	0.7692	0.7407	0.7143	0.6667
2	0.9803	0.9615	0.9426	0.9070	0.8573	0.8264	0.7972	0.7561	0.7182	0.6944	0.6400	0.5917	0.5487	0.5102	0.4444
3	0.9706	0.9423	0.9151	0.8638	0.7938	0.7513	0.7118	0.6575	0.6086	0.5787	0.5120	0.4552	0.4064	0.3644	0.2963
4	0.9610	0.9238	0.8885	0.8227	0.7350	0.6830	0.6355	0.5717	0.5158	0.4822	0.4096	0.3501	0.3011	0.2603	0.1975
5	0.9515	0.9057	0.8626	0.7835	0.6806	0.6209	0.5674	0.4972	0.4371	0.4019	0.3277	0.2693	0.2330	0.1859	0.1317
6	0.9420	0.8880	0.8375	0.7462	0.6302	0.5645	0.5066	0.4323	0.3704	0.3349	0.2621	0.2072	0.1652	0.1328	0.0878
7	0.9327	0.8706	0.8131	0.7107	0.5835	0.5132	0.4523	0.3759	0.3139	0.2791	0.2097	0.1594	0.1224	0.0949	0.0585
8	0.9235	0.8535	0.7894	0.6768	0.5403	0.4665	0.4039	0.3269	0.2660	0.2326	0.1678	0.1226	0.0906	0.0678	0.0390
9	0.9143	0.8368	0.7664	0.6446	0.5002	0.4241	0.3606	0.2843	0.2255	0.1938	0.1342	0.0943	0.0671	0.0484	0.0260
10	0.9053	0.8203	0.7441	0.6139	0.4632	0.3855	0.3220	0.2472	0.1911	0.1615	0.1074	0.0725	0.0497	0.0346	0.0173
11	0.8963	0.8043	0.7224	0.5847	0.4289	0.3505	0.2875	0.2149	0.1619	0.1346	0.0859	0.0558	0.0368	0.0247	0.0116
12	0.8874	0.7885	0.7014	0.5568	0.3871	0.3186	0.2567	0.1869	0.1372	0.1122	0.0687	0.0429	0.0273	0.0176	0.0077
13	0.8787	0.7730	0.6810	0.5303	0.3677	0.2897	0.2292	0.1625	0.1163	0.0935	0.0550	0.0330	0.0202	0.0126	0.0051
14	0.8700	0.7579	0.6611	0.5051	0.3405	0.2633	0.2046	0.1413	0.0985	0.0779	0.0440	0.0254	0.0150	0.0090	0.0034
15	0.8613	0.7430	0.6419	0.4810	0.3132	0.2394	0.1827	0.1229	0.0835	0.0649	0.0352	0.0195	0.111	0.0064	0.0023
16	0.8528	0.7284	0.6232	0.4581	0.2919	0.2176	0.1631	0.1069	0.0708	0.0541	0.0281	0.0150	0.0082	0.0046	0.0015
17	0.8444	0.7142	0.6050	0.4363	0.2703	0.1978	0.1456	0.0929	0.0600	0.0451	0.0225	0.0116	0.0061	0.0033	0.0010
18	0.8360	0.7002	0.5874	0.4155	0.2502	0.1799	0.1300	0.0808	0.0508	0.0376	0.0180	0.0089	0.0045	0.0023	0.0007
19	0.8277	0.6864	0.5703	0.3957	0.2317	0.1635	0.1161	0.0703	0.0431	0.0313	0.0144	0.0068	0.0033	0.0017	0.0005
20	0.8195	0.6730	0.5537	0.3769	0.2145	0.1486	0.1037	0.0611	0.0365	0.0261	0.0115	0.0053	0.0025	0.0012	0.0003

Table B. *Growth at compound interest*

Years (n)	Interest rate (i)														
	0.01(1%)	0.02(2%)	0.03(3%)	0.05(5%)	0.08(8%)	0.10(10%)	0.12(12%)	0.15(15%)	0.18(18%)	0.20(20%)	0.25(25%)	0.30(30%)	0.35(35%)	0.40(40%)	0.50(50%)
1	1.0100	1.0200	1.0350	1.0500	1.0800	1.1000	1.1200	1.1500	1.1800	1.2000	1.2500	1.3000	1.3500	1.4000	1.5000
2	1.0201	1.0404	1.0609	1.1025	1.1664	1.2100	1.2544	1.3225	1.3924	1.4400	1.5625	1.6900	1.8225	1.9600	2.2500
3	1.0303	1.0612	1.0927	1.1576	1.2597	1.3310	1.4049	1.5209	1.6430	1.7280	1.9531	2.1970	2.4604	2.7440	3.3750
4	1.0406	1.0824	1.1255	1.2155	1.3605	1.4641	1.5735	1.7490	1.9388	2.0736	2.4414	2.8561	3.3215	3.8416	5.0625
5	1.0510	1.1041	1.1593	1.2763	1.4693	1.6105	1.7623	2.0114	2.2878	2.4883	3.0518	3.7129	4.4840	5.3782	7.5938
6	1.0615	1.1262	1.1941	1.3401	1.5869	1.7716	1.9738	2.3131	2.6995	2.9860	3.8147	4.8268	6.0534	7.5295	11.3906
7	1.0721	1.1487	1.2299	1.4071	1.7138	1.9487	2.2107	2.6600	3.1855	3.5832	4.7684	6.2749	8.1721	10.5414	17.0859
8	1.0829	1.1717	1.2668	1.4775	1.8509	2.1436	2.4760	3.0590	3.7589	4.2998	5.9605	8.1573	11.0324	14.7579	25.6289
9	1.0937	1.1951	1.3048	1.5513	1.9990	2.3579	2.7731	3.5179	4.4354	5.1598	7.4506	10.6044	14.8937	20.6610	38.4433
10	1.1046	1.2190	1.3439	1.6289	2.1589	2.5937	3.1058	4.0456	5.2338	6.1917	9.3132	13.7858	20.1066	28.9255	57.6650
11	1.1157	1.2434	1.3842	1.7103	2.3316	2.8531	3.4785	4.6524	6.1759	7.4301	11.6415	17.9216	27.1438	40.4957	86.4976
12	1.1268	1.2682	1.4258	1.7959	2.5182	3.1384	3.8960	5.3502	7.2876	8.9161	14.5520	23.2981	36.6442	56.6939	129.7463
13	1.1381	1.2936	1.4685	1.8856	2.7196	3.4523	4.3635	6.1528	8.5994	10.6993	18.1899	30.2875	49.4697	79.3715	194.6195
14	1.1495	1.3195	1.5126	1.9799	2.9372	3.7975	4.8871	7.0757	10.1472	12.8392	22.7374	39.3738	66.7840	111.1200	291.9292
15	1.1610	1.3459	1.5580	2.0798	3.1722	4.1772	5.4736	8.1371	11.9737	15.4070	28.4217	51.1859	90.1585	155.5681	437.8939
16	1.1726	1.3728	1.6047	2.1829	3.4259	4.5950	6.1304	9.3576	14.1290	18.4884	35.5271	66.5417	121.7140	217.7953	656.8408
17	1.1843	1.4002	1.6528	2.2920	3.7000	5.0545	6.8660	10.7613	16.6722	22.1861	44.4089	86.5041	164.3138	304.9135	985.2612
18	1.1961	1.4282	1.7024	2.4066	3.9960	5.5599	7.6900	12.3754	19.6732	26.6233	55.5111	112.4554	221.8236	426.8789	1477.8919
19	1.2081	1.4568	1.7535	2.5270	4.3157	6.1159	8.6128	14.2318	23.2144	31.9480	69.3889	146.1920	299.4619	597.6304	2216.8378
20	1.2202	1.4859	1.8061	2.6533	4.6601	6.7275	9.6463	16.3665	27.3930	38.3376	86.7362	190.0496	404.2736	836.6826	3325.2567

Table C. Present value of an annuity

Years (*n*) — Interest rate (*i*)

Years (n)	0.01(1%)	0.02(2%)	0.03(3%)	0.05(5%)	0.08(8%)	0.10(10%)	0.12(12%)	0.15(15%)	0.18(18%)	0.20(20%)	0.25(25%)	0.30(30%)	0.35(35%)	0.40(40%)	0.50(50%)
1	0.9901	0.9804	0.9709	0.9524	0.9259	0.9091	0.8929	0.8696	0.8475	0.8333	0.8000	0.7692	0.7407	0.7143	0.6667
2	1.9704	1.9416	1.9135	1.8594	1.7833	1.7355	1.6900	1.6257	1.5656	1.5278	1.4400	1.3609	1.2894	1.2245	1.1111
3	2.9410	2.8839	2.8286	2.7232	2.5771	2.4868	2.4018	2.2832	2.1743	2.1065	1.9520	1.8161	1.6959	1.5889	1.4074
4	3.9020	3.8077	3.7171	3.5459	3.3121	3.1699	3.0373	2.8550	2.6901	2.5887	2.3616	2.1662	1.9969	1.8492	1.6049
5	4.8534	4.7135	4.5797	4.3295	3.9927	3.7908	3.6048	3.3522	3.1272	2.9906	2.6893	2.4356	2.2199	2.0352	1.7366
6	5.7955	5.6014	5.4172	5.0757	4.6229	4.3553	4.1114	3.7845	3.4976	3.3255	2.9514	2.6427	2.3852	2.1680	1.8244
7	6.7282	6.4720	6.2303	5.7864	5.2064	4.8684	4.5638	4.1604	3.8115	3.6045	3.1613	2.8021	2.5078	2.2628	1.8829
8	7.6517	7.3255	7.0197	6.4632	5.7466	5.3349	4.9676	4.4873	4.0776	3.8372	3.3290	2.9245	2.5982	2.3306	1.9219
9	8.5660	8.1622	7.7861	7.1078	6.2469	5.7590	5.3282	4.7716	4.3030	4.0310	3.4631	3.0190	2.6653	2.3790	1.9480
10	9.4713	8.9826	8.5302	7.7217	6.7101	6.1446	5.6502	5.0188	4.4941	4.1925	3.5705	3.0915	2.7150	2.4136	1.9653
11	10.3676	9.7868	9.2526	8.3064	7.1390	6.4851	5.9377	5.2337	4.6560	4.3271	3.6564	3.1473	2.7519	2.4383	1.9768
12	11.2551	10.5753	9.9540	8.8632	7.5361	6.8137	6.1944	5.4206	4.7932	4.4392	3.7251	3.1903	2.7792	2.4559	1.9846
13	12.1337	11.3484	10.6349	9.3936	7.9038	7.1034	6.4235	5.5831	4.9095	4.5327	3.7801	3.2232	2.7994	2.4685	1.9897
14	13.0037	12.1062	11.2961	9.8986	8.2442	7.3667	6.6282	5.7245	5.0081	4.6106	3.8241	3.2487	2.8153	2.4775	1.9931
15	13.8650	12.8493	11.9397	10.3797	8.5595	7.6061	6.8109	5.8474	5.0916	4.6755	3.8592	3.2682	2.8255	2.4839	1.9954
16	14.7179	13.5777	12.5611	10.8378	8.8514	7.8237	6.9740	5.9542	5.1623	4.7296	3.8874	3.2832	2.8337	2.4885	1.9970
17	15.5622	14.2919	13.1661	11.2741	9.1216	8.0215	7.1196	6.0472	5.2223	4.7746	3.9099	3.2948	2.8398	2.4918	1.9980
18	16.3983	14.9920	13.7535	11.6896	9.3719	8.2014	7.2497	6.1280	5.2732	4.8122	3.9279	3.3037	2.8443	2.4941	1.9986
19	17.2260	15.6785	14.3238	12.0853	9.6036	8.3649	7.3658	6.1982	5.3162	4.8435	3.9424	3.3105	2.8476	2.4958	1.9990
20	18.0455	16.3514	14.8775	12.4622	9.8181	8.5136	7.4694	6.2593	5.3527	4.8696	3.9539	3.3158	2.8501	2.4970	1.9993

Table D. *Terminal value of a unit annuity*

Years (n)	Interest rate (i)														
	0.01 (1%)	0.02(2%)	0.03(3%)	0.05(5%)	0.08(8%)	0.10(10%)	0.12(12%)	0.15(15%)	0.18(18%)	0.20(20%)	0.25(25%)	0.30(30%)	0.35(35%)	0.40(40%)	0.50(50%)
1	1.0000	1.0000	1.0000	1.0000	1.0000	1.0000	1.0000	1.0000	1.000	1.0000	1.0000	1.0000	1.0000	1.0000	1.0000
2	2.1000	2.0200	2.0300	2.0500	2.0800	2.1000	2.1200	2.1500	2.1800	2.2000	2.2500	2.3000	2.3500	2.4000	2.5000
3	3.0301	3.0604	3.0909	3.1525	3.2464	3.3100	3.3744	3.4725	3.5724	3.6400	3.8125	3.9900	4.1725	4.3600	4.7500
4	4.0604	4.1216	4.1836	4.3101	4.5061	4.6410	4.7793	4.9934	5.2154	5.3680	5.7656	6.1870	6.6329	7.1040	8.1250
5	5.1010	5.2040	5.3091	5.5256	5.8666	6.1051	6.3051	6.7424	7.1542	7.4416	8.2070	9.0431	9.9544	10.9456	13.1875
6	6.1520	6.3081	6.4684	6.8019	7.3359	7.7156	8.1152	8.7537	9.4420	9.9299	11.2588	12.7560	14.4384	16.3238	20.7813
7	7.2135	7.4343	7.6625	8.1420	8.9228	9.4872	10.0890	11.0668	12.1415	12.9159	15.0735	17.5828	20.4919	23.8534	32.1719
8	8.2857	8.5830	8.8923	9.5491	10.6366	11.4359	12.2997	13.7268	15.3270	16.4991	19.8412	23.8577	28.6640	34.3947	49.2578
9	9.3685	9.7546	10.1591	11.0266	12.4876	13.5795	14.7756	16.7858	19.0858	20.7989	25.8023	32.0145	39.6964	49.1526	74.8867
10	10.4622	10.9497	11.4639	12.5779	14.4866	15.9374	17.5487	20.3037	23.5213	25.9587	33.2523	42.6195	54.5902	69.8137	113.3301
11	11.5668	12.1687	12.8078	14.2068	16.6455	18.5312	20.6546	24.3493	28.7551	32.1504	42.5661	56.4053	74.6967	98.7391	170.9951
12	12.6825	13.4121	14.1920	15.9171	18.9771	21.3843	24.1331	29.0017	34.9311	39.5805	54.2077	74.3269	101.8406	139.2348	257.4923
13	13.8093	14.6803	15.6178	17.7130	21.4953	24.5227	28.0291	34.3519	42.2187	48.4966	68.7596	97.6250	138.4848	195.9287	387.2390
14	14.9474	15.9739	17.0860	19.5986	24.2149	27.9750	32.3926	40.5047	50.8180	59.1959	86.9495	127.9125	187.9544	275.3002	581.8585
15	16.0969	17.2934	18.5989	21.5786	27.1521	31.7725	37.2797	47.5804	60.9653	72.0351	109.6868	167.2863	254.7385	386.4202	873.7878
16	17.2579	18.6393	20.1569	23.6575	30.3243	35.9497	42.7533	55.7175	72.9390	87.4421	138.1085	218.4722	344.8969	541.9883	1311.6817
17	18.4304	20.0121	21.7616	25.8404	33.7502	40.5447	48.8837	65.0751	87.0680	105.9305	173.6357	285.0139	466.6109	759.7837	1968.5225
18	19.6147	21.4123	23.4144	28.1324	37.4502	45.5992	55.7497	75.8363	103.7403	128.1167	218.0446	371.5180	630.9247	1064.6971	2953.7838
19	20.8109	22.8406	25.1169	30.5390	41.4463	51.1591	63.4397	88.2118	123.4135	154.7400	273.5558	483.9734	852.7483	1491.5759	4431.6756
20	22.0190	24.2974	26.8704	33.0660	45.7620	57.2750	72.0524	102.4436	146.6280	186.6880	342.9445	630.1655	1152.2103	2089.2064	6648.5135

Table E. *Annuity whose terminal value is 1*

Years (n)	0.01(1%)	0.02(2%)	0.03(3%)	0.05(5%)	0.08(8%)	0.10(10%)	0.12(12%)	0.15(15%)	0.18(18%)	0.20(20%)	0.25(25%)	0.30(30%)	0.35(35%)	0.40(40%)	0.50(50%)
1	1.0000	1.0000	1.0000	1.0000	1.0000	1.0000	1.0000	1.0000	1.0000	1.0000	1.0000	1.0000	1.0000	1.0000	1.0000
2	0.4975	0.4950	0.4926	0.4878	0.4808	0.4762	0.4717	0.4651	0.4587	0.4545	0.4444	0.4348	0.4255	0.4167	0.4000
3	0.3300	0.3267	0.3235	0.3172	0.3080	0.3021	0.2963	0.2880	0.2799	0.2747	0.2623	0.2506	0.2397	0.2294	0.2105
4	0.2463	0.2426	0.2390	0.2320	0.2219	0.2155	0.2092	0.2003	0.1917	0.1863	0.1734	0.1616	0.1508	0.1408	0.1231
5	0.1960	0.1921	0.1883	0.1810	0.1705	0.1638	0.1574	0.1483	0.1398	0.1344	0.1218	0.1106	0.1005	0.0914	0.0758
6	0.1625	0.1585	0.1546	0.1470	0.1363	0.1296	0.1232	0.1142	0.1059	0.1007	0.0888	0.0784	0.0693	0.0613	0.0481
7	0.1386	0.1345	0.1305	0.1227	0.1121	0.1054	0.0991	0.0904	0.0824	0.0774	0.0663	0.0569	0.0488	0.0419	0.0311
8	0.1201	0.1165	0.1124	0.1047	0.0940	0.0874	0.0813	0.0728	0.0652	0.0606	0.0504	0.0419	0.0349	0.0291	0.0203
9	0.1067	0.1025	0.0984	0.0907	0.0801	0.0736	0.0677	0.0596	0.0524	0.0481	0.3876	0.0312	0.0252	0.0203	0.0134
10	0.0956	0.0913	0.0872	0.0795	0.0690	0.0627	0.0570	0.0492	0.0425	0.0385	0.0301	0.0235	0.0183	0.0143	0.0088
11	0.0864	0.0822	0.0781	0.0704	0.0601	0.0540	0.0484	0.0411	0.0348	0.0311	0.0235	0.0177	0.0134	0.0101	0.0058
12	0.0788	0.0745	0.0705	0.0628	0.0527	0.0468	0.0414	0.0345	0.0286	0.0253	0.0184	0.0135	0.0098	0.0072	0.0039
13	0.0724	0.0681	0.0640	0.0564	0.0465	0.0408	0.0357	0.0291	0.0237	0.0206	0.1454	0.0124	0.0072	0.0051	0.0026
14	0.0669	0.0626	0.0585	0.0510	0.0413	0.0357	0.0309	0.0247	0.0197	0.0169	0.0115	0.0078	0.0053	0.0036	0.0017
15	0.0621	0.0578	0.0538	0.0463	0.0368	0.0315	0.0268	0.0210	0.0164	0.0139	0.0091	0.0059	0.0039	0.0026	0.0011
16	0.0579	0.0536	0.0496	0.0427	0.0330	0.0278	0.0234	0.0179	0.0137	0.0114	0.0072	0.0046	0.0029	0.0018	0.0008
17	0.0543	0.0500	0.0459	0.0387	0.0296	0.0247	0.0205	0.0154	0.0115	0.0094	0.0058	0.0035	0.0021	0.0013	0.0005
18	0.0510	0.0467	0.0427	0.0355	0.0267	0.0219	0.0179	0.0132	0.0096	0.0078	0.0046	0.0027	0.0016	0.0009	0.0003
19	0.0480	0.0438	0.0398	0.0327	0.0241	0.0195	0.0158	0.0113	0.0081	0.0065	0.0036	0.0021	0.0012	0.0007	0.0002
20	0.0454	0.0411	0.0372	0.0302	0.0218	0.0175	0.0139	0.0098	0.0068	0.0054	0.0029	0.0016	0.0009	0.0005	0.0002

Interest rate (i)

Appendix 2: Metric conversion

The conversion of these measurements to metric units is based on the International System of Units, known as S.I. The degree of accuracy used in the following tables will be quite unnecessary for most conversions; asking for 76.2 millimetre nails or a length of rope 3.048 metres long at the local hardware store may bewilder the man behind the counter.

Mr A. Gates, of the University of New England, has prepared these tables from basic data supplied by the Metric Conversion Board (Australia).

The basic units

Unit	Imperial Name	Abbreviation	Metric Name	Abbreviation
Length	foot	ft	metre	m
Mass	pound	lb.	gram	g
Area	square foot	sq. ft	square metre	m^2
Volume	cubic foot	c. ft	cubic metre	m^3
Volume (fluids)	gallon	gal.	litre	l
Time	second	sec.	second	s
Force	pound force	lbf	newton	N
Pressure	pound per square inch	psi	pascal	Pa
Energy	British thermal unit	Btu	joule	J
Power	horsepower	hp	watt	W
Angle	degree	°	radian	rad

Prefixes for multiples and fractions of basic SI units

Fractions

(0.1)	deci (d)
(0.01)	centi (c)
(0.001)	milli (m)
(0.000 001)	micro (μ)
(0.000 000 001)	nano (n)

Multiples

10	deka (d)
100	hecto (h)
1000	kilo (k)
1 000 000	mega (M)
1 000 000 000	giga (G)

When writing numbers in the imperial system, the practice is to space each third digit from the decimal point (thousand and thousandths) with a comma, – 3,141,646.347,2.

The S.I. system does not use commas – you just leave a space thus – 3 141 646.347 2.

Length

Imperial	Metric
1 foot (ft) =	0.304 8 metres
1 yard (yd) =	0.914 4 m
1 mile =	1.609 4 km

Metric	Imperial
1 mm =	0.039 37 in
10 mm (1 centimetre) =	0.393 70 in
100 mm (1 decimetre) =	3.937 01 in
1000 mm (1 metre) =	39.370 1 in
1 metre (m) =	3.280 84 ft
1 m =	1.093 613 yd
1000 m (1 km) =	0.621 371 miles

Mass

Imperial	Metric (SI)
1 ounce (oz) =	28.349 52 grams (g)
1 pound (lb) =	453.592 37 g
1 pound (lb) =	0.453 592 kilograms (kg)
1 short ton (2000 lb) =	907.184 74 kg
1 ton (2240 lb) =	1.016 047 tonnes (t)

Metric (SI)	Imperial
1 g =	0.035 274 oz
1 kg =	2.204 623 lb
1 t =	2204.622 6 lb
1 t =	1.102 311 short tons (2000 lb)
1 t =	0.984 206 ton (2240 lb)

Area

Imperial	Metric
1 square inch (in²) =	6.451 6 square centimetres (cm²)
1 square foot (ft²) =	0.092 903 square metres (m²)
1 square yard (yd²) =	0.836 127 square metres (m²)
1 acre =	0.404 686 hectares (ha)
1 square mile =	2.589 988 square kilometres (km²)

Metric	Imperial
1 square centimetre (cm²) =	0.155 square inches (in²)
1 square metre (m²) =	10.763 9 square feet (ft²)
1 square metre (m²) =	1.195 99 square yards (yd²)
1 hectare (ha) =	2.471 05 acres
1 square kilometre (km²) =	0.386 102 square miles

Note: pounds per acre × 1.121 = kg/hectare

Volume

1 cubic inch (in³) =	16.387 1 cubic centimetres (cm³)
1 cubic foot (ft³) =	0.028 317 cubic metre (m³)
1 cubic yard (yd³) =	0.764 555 cubic metre (m³)
1 bushel (bush) =	0.363 69 hectolitre (hl)
1 bushel (bush) =	0.036 369 cubic metre (m³)
1 cubic centimetre (cm³) =	0.061 024 cubic inch (in³)
1 cubic metre (m³) =	35.314 6 cubic feet (ft³)
1 cubic metre (m³) =	1.307 95 cubic yards (yd³)
1 hectolitre (hl) =	2.749 62 bushels (bus)
1 cubic metre (m³) =	27.496 2 bushels (bus)
1 fluid ounce (fl. oz.) =	28.411 millilitres (ml)

1 pint (pt) =	568.261 millilitres (ml)
1 gallon (gal.) =	4.546 09 litres (l)
1 acre inch (ac. in.) =	10.279 hectare millimetres (ha mm)
1 acre foot (ac. ft) =	1233.48 cubic metres (m³) or 1.233 megalitres (Ml)
1 millilitre (ml) =	0.035 2 fluid ounces (fl. oz)
1 litre (l) =	1.759 75 pints (pt)
1 kilolitre (kl) (or cubic metre) =	219.969 gallons (gal.)
1 megalitre (Ml) =	0.8107 acre foot (ac. ft)
1 hectare millimetre (ha mm) =	0.097 286 acre inch (ac. in.)

Density

1 pound per gallon (lb/gal) =	0.099 779 kilograms per litre (kg/l)
1 pound per cubic foot (lb/ft) =	16.081 5 kilograms per cubic metre (kg/m)
1 kilogram per litre (kg/l) =	10.022 1 pounds per gallon (lb/gal.)
1 kilogram per cubic metre (kg/m) =	0.062 428 pounds per cu. ft (lb/ft)

Energy

1 calorie (cal) =	4.186 8 joules (J)
1 kilowatt hour (kWh) =	3.6 megajoules (MJ)

Metric	
1 joule (J) =	0.238 846 calories (cal)
1 megajoule (MJ) =	0.009 479 therm
1 megajoule (MJ) =	0.277 778 kilowatt hour (kWh)

Power

1 horsepower (hp) =	0.745 7 kilowatt (kW)
1 kilowatt (kW) =	1.341 02 horsepower (hp)

Pressure

1 pound per square inch (psi) =	6.89 kilopascals (kPa)
1 kilopascal (kPa) =	0.14 pound per square inch (psi)

Glossary

Here are some of the more common terms used in farm management economics. The number(s) refer to the chapter(s) where each term is described more fully.

Activity – a particular method of producing a commodity. More specific term than *enterprise*, e.g. broiler chickens, second crop rice – 8.

Agistment – grazing one's animals on land controlled by another person, for the payment of a fee in cash or in kind – 12.

Amortise (to kill) – see 'sinking fund' – 13, 14, 15.

Annuity – a sum of money received or spent in equal amounts over a period – 13.

Benefit : cost ratio – the ratio of the present value of cash received over the life of a development project to the present value of costs (payments) – 9, 14.

Budget – a detailed statement of a plan which describes both costs involved and the returns expected – 7, 10.

Budget control – the process of comparing the actual performance of an aspect of farm production against the performance which was expected when the budget was drawn up – 7.

Capital – goods which have not been used up, includes land, equipment, livestock and cash – 4, 5, 6.

Capital investment – money spent on equipment, stock or on improvements which have a life of more than 1 year and which add to the productive capacity of the farm – 5, 6.

Capital gains – increase in the value of capital items due to a rise in their market price – 5, 6.

Cash flow – the movement of money in and out of the hands of an enterprise or individual farmers (see also *net cash flow*) – 7.

Cash flow budget – a budget of the expected cash in (receipts) and cash out (payments) associated with a particular farm plan – 7, 14.

Cast for age (CFA) – a reject old animal that is past its economic life for particular conditions – 12.

Complementary products – where the process of producing product '*X*' also leads to more of product '*Y*' – 4.

Compounding – adding interest to a sum of money at a chosen rate, including interest on the interest accumulated each year, i.e. calculation of the future value of a present sum – 9.

Compound interest rate – the rate of interest used in compounding or discounting – 9.

Contingency allowance – allowance to cover unexpected events, e.g. a drought resulting in severe losses of cattle or crops – 10, 14.

Cropping intensity – the number of years of cropping, multiplied by 100, and divided by the number of years of the rotation. It is expressed as '*R*'. For example, 3 years crop, 7 years fallow = 10 year rotation. Thus, $R = (3 \times 100)/10 = 30 - 11$.

Debt-servicing capacity – annual whole farm net cash flow before deducting interest and loan repayments – 15.

Depreciation – the loss in value of capital equipment as it becomes older – 4, 5.

Depreciation allowance – the sum of money which is deducted from income each year so that funds are available to replace equipment, etc. when it is worn out – 4, 5, 13.

Development budget – a budget used when planning major changes in a farm which will take some time to reach full capacity – 14.

Diminishing returns – the phenomenon that after some point, increases in variable inputs to a production process results in smaller and smaller increases in total output. The principle of diminishing returns indicates that variable input should be added to the production process so long as the extra return exceeds the extra cost, and the maximum total profit is at the point where extra return equals the extra cost – 4.

Discounting – calculation of the present value of a future sum – 9.

Discounting factor – the value by which a future cash flow must be multiplied to calculate its present value – 9.

Enterprise – the production of a particular commodity or group of related commodities. A general term, e.g. maize – 2, 7.

Equity – assets minus liabilities – 6, 15.

Equity per cent – farm equity capital as a percentage of total farm capital, i.e.

$$\frac{\text{assets minus liabilities}}{\text{assets}} \times \frac{100}{1}$$

– 6, 15.

Extra (marginal) capital – additional capital invested in a farm with a view to increasing profit, e.g. clearing land, building a small dam for irrigation – 10, 14.

Extra (marginal) cost – the cost of adding one more unit of input to a production process or the cost of producing one extra unit of return – 4.

Extra (marginal) return – the extra return (revenue) which results from adding one more unit of input – 10.

Finance budget – a budget showing the borrowings which are needed and interest and principal repayments – 15.

Fixed capital – land, buildings, bores, irrigation equipment etc., which cannot easily be moved – 4, 5.

Fixed (overhead) costs – costs which must be met and are not affected by the size of the activities in the farm operations – 5.

Gross income – the total value of a farm activity, i.e. during a production period (usually 1 year), whether the product is sold, consumed or stored – 5.

Gross margin – gross income minus variable costs – 8.

Gross margins planning – a procedure whereby activities are selected on the basis of the gross margin from a unit of only one key constraint, usually labour or land – 8.

Growth 'profit' – increase in equity or savings – 15.

Inflation – an increase in the supply of money in relation to the goods and services available and, in consequence, a decline in its value – 9, 15.

Input – any resource used in production, e.g. land, labour or capital – 4.

Internal rate of return – the discount rate at which the present value of a future income from a project equals the present value of total expenditure (capital and annual costs) on the project – 9, 14.

Interest – the cost of 'hiring' money from a lender – 15.

Interest (flat) – a way of calculating interest in hire purchase and lease agreements. The interest is charged, over the period of the loan, on the amount originally borrowed, even though some of the loan has been repaid. It is nearly twice as expensive as simple interest – 15.

Interest (simple) – the hiring charge of borrowed money, calculated on the amount owed, at any one time – 15.

Investment appraisal – an evaluation of the profitability of a farm development plan – 9, 14.

Leverage (gearing) – the ratio of debts to equity – 15.

Linear programming – a mathematical, computer-based, farm planning technique which can be used to determine the combination of activities which maximizes total 'profit', or minimises costs – 11.

Liquidity – the ease with which assets can be converted into cash. For example, animals and jewellery are liquid assets, land less so – 15.

Livestock feed budget – a budget comparing feed requirements of livestock with the feed available – 12.

Livestock gross income – the value of livestock production in the form of animals and produce, adjusted for inventory changes – 8, 12.

Marginal – economists' word for 'extra' or 'added'. The principle of marginality refers to the profit-maximising level of operation where the marginal revenue from production equals the marginal cost of production – 4.

Marginal cost – the extra cost incurred of adding one more unit of input to the production process – 4.

Marginal product – the change in output arising from using an extra unit of an input – 4.

Marginal revenue – the extra net income obtained from adding one more unit of input – 4.

Net cash flow – the difference between the money received and the money spent in any one period (week, month, year) – 7.

Net present value (NPV) – the sum of the discounted values of the future income and costs associated with a given farm project or plan – 9.

Net worth (equity) – the value of total assets that the farmer owns minus the value of his total liabilities – 6, 15.

Nominal dollar – the face value of a dollar at any point in time. Nominal money values are expressed in terms of the actual present day price of things one has to buy – 9.

Operating costs – variable costs plus overhead (fixed) costs – 5, 6.

Operating profit – gross income minus variable and operating overhead costs – 5, 6.

Opportunity cost – the opportunity cost of a farm management decision is the amount of money which is given up by choosing one alternative rather than another – 4, 5.

Output – the goods or products, e.g. food and fibre, which come from a production process – 4.

Overhead (fixed) costs – costs which do not vary greatly as the level of production or mixture of activities changes, but which must be met each year – 5, 6.

Parameter – any factor which has an important effect on 'profit' (yield, price, hectarage, direct cost) – 10.

Parametric budget – a planning technique which takes varying prices and yields into account – 10.

Partial budget – a budget which shows the expected extra expenses and extra returns for a change contemplated in the farm programme – 10.

Payment in kind – Non-cash payment for a service received (e.g. labour). Payment may be in the form of, for instance, food, an equivalent amount of labour, or in machinery services – 5.

Principle of comparative advantage – an economic principle recognising that various crops and livestock should be produced in areas where the physical and other resources most favour such production – 4, 17.

Production (or response) function – the relationship between the level of inputs and the level of output for a production process – 4.

Profit – a general term indicating some kind of surplus from the years farming operations. There are many definitions of profit – 6.

Real dollar – the value of a dollar over time with the effect of changes in the purchasing power removed; thus real values are comparable (cf. nominal dollar) – 9.

Receipts – money received by the farmer from either farm or non-farm sources – 5.

Resource – a factor of production. Commonly classified under labour, land, capital, raw materials – 4.

Response (production) function – the relationship between the amount of variable inputs (e.g. weedings) applied to a fixed resource (land) and the output (crop) which results from the inputs – 4.

Return on extra (marginal) capital – the extra 'profit' resulting from investing extra capital in the farm, expressed as a percentage – 5, 10, 14.

Return to total capital – the annual operating profit expressed as a percentage of total capital. This figure does not take account of interest, loan repayments, living costs, new capital and tax payments – 6.

Risk – a situation with uncertain outcome, e.g. next year's farm income – 10.

Self-replacing – a system of animal production where part of the female progeny is retained each year to replace the older females sold – 12.

Sensitivity testing – checking the effect on a planned outcome of a change in one of the factors (parameters, coefficients) contributing to that outcome – 10.

Simplified programming – a technique, which does not need access to a computer, to help plan mixed-cropping farms – 11.

Sinking fund – An annual payment which is made in order to allow for a periodic payment of a debt or replacement of an asset.

Sinking fund factor – an amount of money received or set aside in equal sums, which will grow to a final, specified, value over a period – 13, Appendix 1.

Stock equivalents – units used in livestock feed budgeting whereby the energy needs of different categories of livestock are expressed in terms of one type of livestock, e.g. livestock months, tropical livestock units – 12.

Supplementary products – where producing more of product 'X' does not reduce the output of product 'Y' – 4.

Variable costs (also direct costs) – costs which change according to the size of the activity, e.g. fuel, seed – 5.

Whole farm budget – budget showing the expected outcomes of a farm plan, as it affects the entire farm's 'profitability' – 6.

Whole farm planning – planning for the entire farm, as distinct from partial budget planning – 6, 10.

Index